Wireless Networks

Series Editor

The purpose of Springer's Wireless Networks book series is to establish the state of the art and set the course for future research and development in wireless communication networks. The scope of this series includes not only all aspects of wireless networks (including cellular networks, WiFi, sensor networks, and vehicular networks), but related areas such as cloud computing and big data. The series serves as a central source of references for wireless networks research and development. It aims to publish thorough and cohesive overviews on specific topics in wireless networks, as well as works that are larger in scope than survey articles and that contain more detailed background information. The series also provides coverage of advanced and timely topics worthy of monographs, contributed volumes, textbooks and handbooks.

** Indexing: Wireless Networks is indexed in EBSCO databases and DPLB **

More information about this series at http://www.springer.com/series/14180

Weihua Zhuang • Kaige Qu

Dynamic Resource Management in Service-Oriented Core Networks

Springer

Weihua Zhuang
Department of Electrical and Computer Engineering
University of Waterloo
Waterloo, ON, Canada

Kaige Qu
Department of Electrical and Computer Engineering
University of Waterloo
Waterloo, ON, Canada

ISSN 2366-1186 ISSN 2366-1445 (electronic)
Wireless Networks
ISBN 978-3-030-87138-3 ISBN 978-3-030-87136-9 (eBook)
https://doi.org/10.1007/978-3-030-87136-9

This Springer imprint is published by the registered company Springer Nature Switzerland AG
The registered company address is: Gewerbestrasse 11, 6330 Cham, Switzerland

Preface

The service-oriented fifth-generation (5G) and beyond core networks are featured by customized network services with differentiated quality-of-service (QoS) requirements, which can be provisioned through network slicing enabled by the software-defined networking (SDN) and network function virtualization (NFV) paradigms. Multiple network services are embedded in a common physical infrastructure, generating service-customized network slices (also referred to as virtual networks). Each network slice supports a composite service via virtual network function (VNF) chaining, with dedicated packet processing functionality at each VNF. For a network slice with a target traffic load, the end-to-end (E2E) service delivery is enabled by VNF placement at NFV nodes (e.g., data centers and commodity servers) and traffic routing among corresponding NFV nodes, with static resource allocations. When data traffic actually enters the network, the traffic load is dynamic and can deviate from the target value, potentially leading to QoS performance degradation and network congestion. There are traffic dynamics in different time granularities. For example, the traffic statistics (e.g., mean and variance) can be non-stationary and can experience significant changes in a coarse time granularity, e.g., larger than 30 min, which are usually predictable. Within a long time duration with stationary traffic statistics, there are traffic dynamics in small timescales, e.g., less than 1 ms, which are usually highly bursty and unpredictable. To provide continuous QoS performance guarantee and ensure efficient and fair operation of the virtual networks over time, it is essential to develop dynamic resource management schemes for the embedded services experiencing traffic dynamics in different time granularities during virtual network operation.

This book provides a timely and comprehensive study of dynamic resource management for network slicing in service-oriented 5G and beyond core networks, from the perspective of developing efficient computing resource provisioning and scheduling solutions to guarantee consistent service performance in terms of E2E delay guarantee. Queueing theory is used in system modeling, and different techniques including optimization and machine learning are applied to solving the dynamic resource management problems. We capture the heterogeneity between computing and communication resources and investigate dynamic computing resource provisioning and scheduling for embedded delay-sensitive services. Based on a simplified M/M/1 queueing model with Poisson traffic arrivals, an optimization model for flow migration is developed to accommodate the large-timescale changes in the average traffic rates with average E2E delay guarantee while addressing a trade-off between load balancing and flow migration overhead. To overcome the limitations of Poisson traffic model, we develop a machine learning approach for dynamic VNF resource scaling and migration. The new solution captures the inherent traffic patterns in a real-world traffic trace with non-stationary traffic statistics in large timescale, predicts resource demands for VNF resource scaling, and triggers adaptive VNF migration decisions, to achieve load balancing, migration cost reduction, and resource overloading penalty suppression in the long run. Both supervised and unsupervised machine learning tools are investigated for dynamic resource management. To accommodate the traffic dynamics in small time granularities, we develop a dynamic VNF scheduling scheme to coordinate the scheduling among VNFs of multiple services, which achieves network utility maximization with delay guarantee for each service.

Finally, after concluding remarks on the research work, we identify further research directions in the dynamic resource management.

Waterloo, ON, Canada Weihua Zhuang

Waterloo, ON, Canada Kaige Qu
July 2021

Contents

Acronyms

5G	Fifth generation
BOCPD	Bayesian online change point detection
CFS	Completely fair scheduler
CPU	Central processing unit
DQN	Deep Q-network
E2E	End-to-end
eMBB	Enhanced mobile broadband
fBm	Fractional Brownian motion
GP	Gaussian process
GPR	Gaussian process regression
GPS	Generalized processor sharing
HFM	Hybrid flow migration
HTTP	Hypertext transfer protocol
IDS	Intrusion detection system
InP	Infrastructure provider
LBFM	Load balancing flow migration
LRD	Long-range dependence
MANO	Management and orchestration
MDP	Markov decision process
MIQCP	Mixed integer quadratically constrained programming
mMTC	Massive machine-type communication
MMPP	Markov-modulated Poisson process
MOFM	Minimum overhead flow migration
M2M	Machine-to-machine
NAT	Network address translator
NFV	Network function virtualization
NFVI	NFV infrastructure
NFVO	NFV orchestrator
NSO	Network service orchestrator
OS	Operating system
QoS	Quality-of-service

QQ	Quantile–quantile
RL	Reinforcement learning
RLS	Real time scheduler
RO	Resource orchestrator
SDN	Software defined networking
SFC	Service function chain
SLA	Service-level agreement
TE	Traffic engineering
uRLLC	Ultra reliable low latency communication
VIM	Virtualized infrastructure manager
VM	Virtual machine
VNF	Virtual network function
VNFM	VNF manager
VR	Virtual reality

Chapter 1
Introduction

1.1 Service-Oriented Core Networks

The service-oriented fifth generation (5G) and beyond networks will support new use cases and diverse services with multi-dimensional performance requirements [1, 2]. There are three typical 5G use case categories: enhanced mobile broadband (eMBB), massive machine-type communication (mMTC), and ultra reliable low latency communication (uRLLC), which demonstrate diversified traffic characteristics and differentiated quality-of-service (QoS) requirements. The eMBB services span several human-centric use cases, which are characterized by high data rates (up to the order of Gbps). Typical applications include 4K/8K ultra-high resolution video streaming, virtual reality, and augmented reality. Both the second and third categories are machine-centric. The mMTC services support massive number of connected devices with low mobility and less stringent QoS requirement in terms of delay and reliability. The uRLLC services have stringent QoS requirements in terms of delay and reliability, typically with an E2E delay requirement in the order of milliseconds and a reliability requirement of higher than 99.999% [3]. Typical uRLLC applications include industry automation, autonomous driving, and remote surgery.

The disparate performance requirements of the three use case families are difficult to be satisfied by the legacy *one-size-fits-all* network architecture. Instead, network slicing is required on a per-service basis, to provide service-level performance guarantees. With network slicing, multiple network slices with diverse performance requirements are embedded over a common physical infrastructure [4–11]. This requires a flexible and programmable network architecture, which facilitates cost-effective and flexible service customization and management. Software defined networking (SDN) and network function virtualization (NFV) are two enabling technologies for such a network architecture, providing abstractions on the plane and layer dimensions respectively [12].

W. Zhuang, K. Qu, *Dynamic Resource Management in Service-Oriented Core Networks*, Wireless Networks, https://doi.org/10.1007/978-3-030-87136-9_1

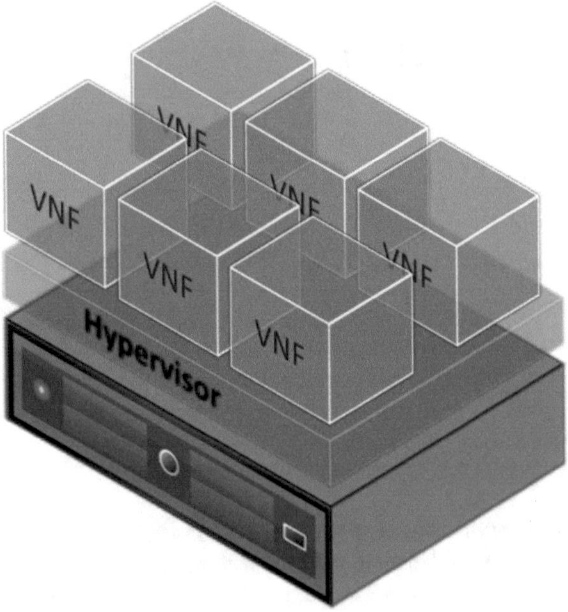

Fig. 1.1 Network function virtualization with hypervisor

Software defined networking (SDN) brings the plane-dimension abstraction by decoupling the data and control planes. The data plane elements, such as SDN-enabled switches, forward traffic according to the control instructions (e.g., forwarding rules) enforced by a logically centralized and programmable control plane. This allows the data plane to be fully abstracted for applications. Moreover, it facilitates a global network view from the perspective of controller, which allows for more efficient resource management. With a global network view and flow awareness brought by SDN, end-to-end (E2E) data delivery paths can be dynamically established by configuring data flow forwarding rules in SDN-enabled switches via southbound protocols such as OpenFlow,[1] and resources are explicitly allocated to different paths.

Traditionally, service providers rely on dedicated hardware middleboxes to realize network functions as in-path packet processing units required by a service, such as intrusion detection system (IDS), network address translator (NAT), firewall, 5G evolved packet core functions, cache, wireless access network optimizer, transcoder, etc. The dedicated hardware middleboxes are fixed and expensive, thereby lacking flexibility in deployment and management, and resulting in high maintenance cost and inefficient resource utilization. Network function virtualization (NFV) provides the layer-dimension abstraction, by abstracting physical resources to virtual resources with a virtualization layer and realizing service-level

[1] OpenFlow is a southbound application programming interface (API) that facilitates the configuration of OpenFlow forwarding tables over the data plane elements by the SDN controller [13].

functionalities [12, 14]. As illustrated in Fig. 1.1, NFV virtualizes network functions from dedicated hardware on NFV nodes such as commodity servers and data centers to software instances, referred to as virtual network functions (VNFs), through a hypervisor. NFV reduces the capital and operational expenditures by enabling on-demand, flexible, and cost-effective VNF placement and elastic VNF capacity scaling.

1.1.1 Network Slicing Framework

Several frameworks have been proposed for SDN-NFV integration, to fully exploit their advantages and provide an integrated architecture with abstractions in both plane and layer dimensions for network slicing and customized service provisioning [12, 15]. Next, we introduce an SDN and NFV enabled network slicing framework from the infrastructure domain to the tenant domain, both of which are managed by an extended NFV management and orchestration (MANO) architecture with SDN integration.

1.1.1.1 Infrastructure Domain

The infrastructure domain is composed of a physical network, a virtual resource pool, and an infrastructure SDN controller, as illustrated in Fig. 1.2.

Physical Network

The physical network is composed by SDN switches and NFV nodes interconnected by physical links. Switches forward traffic from incoming physical links to outgoing physical links. Some switches serve as edge switches for service access. NFV nodes, such as data centers (DCs) and commodity servers, have the capabilities of both data forwarding and processing. The physical network contains a physical resource pool, consisting of computing resources at NFV nodes and transmission resources on physical links. A path in the physical network, i.e., a physical path, consists of a series of physical links and SDN switches between one NFV node and one edge switch or between two NFV nodes. A central orchestrator determines the maximum transmission rate that a physical route can handle.

Virtual Resource Pool

There can be multiple physical paths with pre-allocated transmission resources between two end points in the physical network, such as two NFV nodes or one NFV node and one edge switch, which can be logically abstracted to a virtual

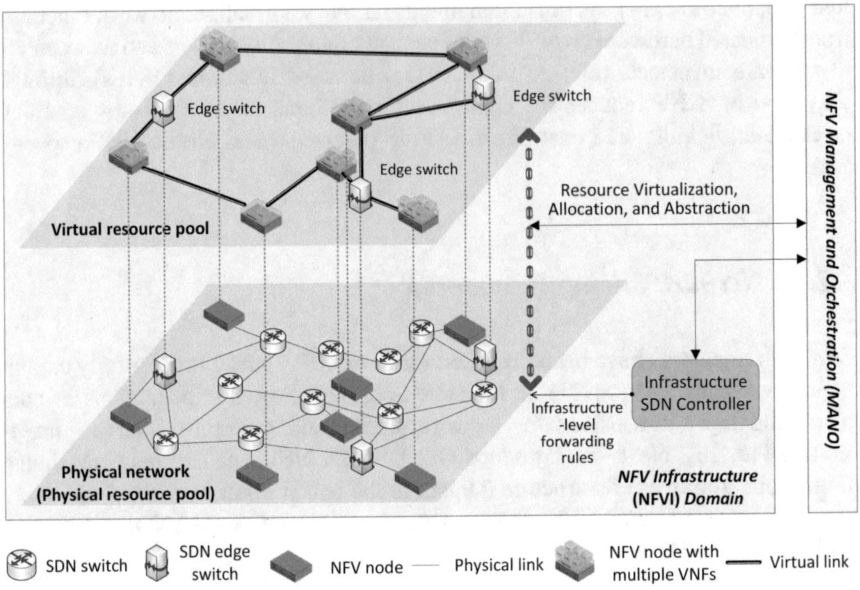

Fig. 1.2 Infrastructure domain

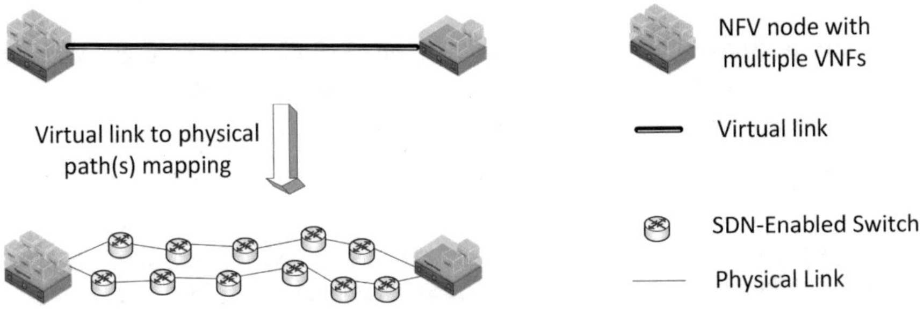

Fig. 1.3 An illustration of virtual link mapping to two physical paths between two NFV nodes

link between the two end points. As illustrated in Fig. 1.3, a virtual link between two different NFV nodes corresponds to a logical abstraction of two physical paths between the two NFV nodes. The maximum transmission rate supported by a virtual link is the aggregate maximum transmission rate over all its underlying physical paths. Transmission resources on virtual links are seen as virtual resources, since the mapping between virtual links and physical paths is transparent to service flows traversing the virtual links. With network function virtualization, the computing resources at NFV nodes are virtualized and distributed among several VNFs through a virtualization hypervisor. Hence, a virtual resource pool containing virtual resources for transmission and computing can be abstracted from the physical resource pool, which makes both SDN switches and physical links fully transparent to service flows.

A path in the virtual resource pool, i.e., a virtual path, is composed of a series of virtual links and NFV nodes between two edge switches. It is possible that the virtual resource pool is not a fully connected graph, i.e., not every two NFV nodes or edge switches are directly connected by a virtual link. The topology of the virtual resource pool can be flexibly updated, via scaling up/down transmission resources on existing virtual links, removing existing virtual links, and finding physical paths with sufficient transmission resources for new virtual links, which enables flexible virtual link provisioning among the fixed NFV nodes.

The resource virtualization, allocation, and abstraction in the infrastructure domain are managed by the NFV MANO architecture. The mapping between virtual links and physical paths is determined by a central orchestrator, maintained by a virtual infrastructure manager (VIM), and enforced by an infrastructure SDN controller.

Infrastructure SDN Controller

With SDN, packet forwarding rules are configured in SDN switches by an infrastructure SDN controller to route traffic flows through a physical path. For virtual link provisioning, the infrastructure SDN controller is responsible for (1) configuring forwarding rules on physical paths associated with each virtual link, and (2) enforcing traffic splitting ratios among corresponding physical paths for each virtual link. When a topology update for the virtual resource pool is required, the infrastructure SDN controller is responsible for (re-)configuring forwarding rules on physical paths for the scaled and new virtual links, and removing those associated with the obsolete virtual links.

1.1.1.2 Tenant Domain

The tenant domain is composed of services and a tenant SDN controller, as illustrated in Fig. 1.4.

Services

A tenant such as a service provider requests network services in the form of service function chains (SFCs). An SFC consists of multiple VNFs in a predefined order, to fulfill a composite service with certain processing and transmission resource demands, based on service level agreements (SLAs) negotiated with an infrastructure provider (InP). Each VNF supports a dedicated packet processing functionality. Hence, a network service can provide customized packet processing functionalities in addition to the traditional transmission connectivity to a group of end users. In an SFC, there are two levels of connectivity, i.e., service-level and infrastructure-level. The service-level connectivity requires VNFs be chained

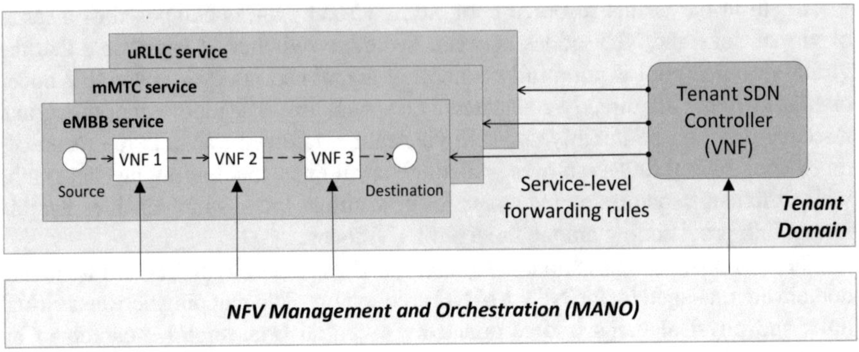

Fig. 1.4 Tenant domain

in a predefined order between destination nodes (fixed at edge switches) and the source, to facilitate the E2E service delivery. An SFC is mapped to a virtual route between the source and destination nodes to enable service-level connectivity. For two neighboring VNFs in an SFC, packets processed by the upstream VNF are transmitted to the downstream VNF, generating traffic between consecutive VNFs, i.e., inter-VNF subflows. The infrastructure-level connectivity requires that each subflow be routed over at least one physical path, if its upstream and downstream VNFs are not co-located. Each subflow is mapped to a virtual link, which is provisioned via the infrastructure SDN controller, to achieve infrastructure-level connection.

Tenant SDN Controller

The tenant SDN controller configures service-level forwarding rules at edge switches and NFV nodes to guide packets belonging to an aggregate traffic flow traversing an SFC (i.e., an SFC flow or a service flow) through a virtual path, thus enabling the service-level connectivity. In the presence of traffic variations, an SFC flow can be rerouted to an alternative virtual path via the tenant SDN controller, according to flow migration decisions made by a central orchestrator in the NFV MANO architecture.

1.1.1.3 SDN-NFV Integration

An NFV management and orchestration (MANO) architecture manages the life cycle of network functions, services, and their constituent resources in a common NFV infrastructure (NFVI) [5]. The architecture is extended with SDN integration to realize service function chaining. The main functional blocks in the architecture

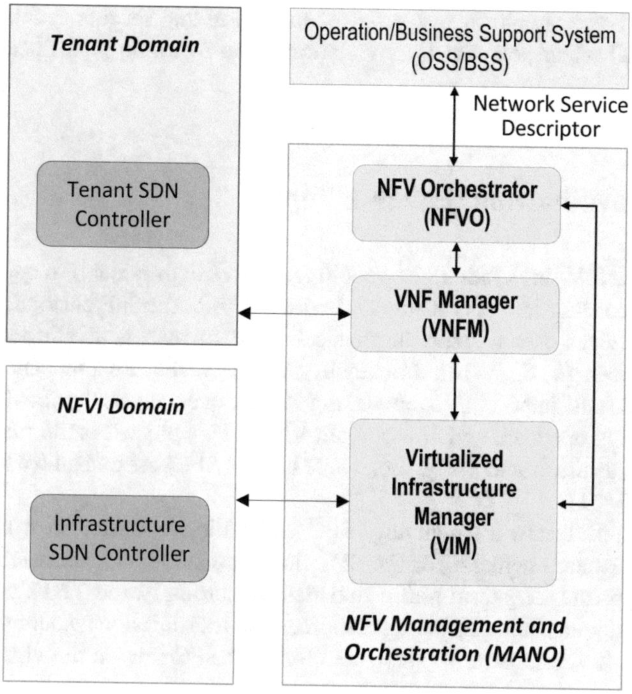

Fig. 1.5 An extended NFV MANO architecture with SDN integration

and their interactions with the tenant and infrastructure SDN controllers are illustrated in Fig. 1.5 and introduced as follows.

- *Virtualized infrastructure manager (VIM)* is responsible for managing resources in the NFVI. Specifically, the VIM deals with resource virtualization and allocation, and maintains the mapping between the virtual resource pool and physical resource pool. The VIM is also in charge of virtual link provisioning via an infrastructure SDN controller;
- *VNF manager (VNFM)* is in charge of the life cycle management of VNFs, including instantiation, configuration, and scaling. In addition to VNFs serving as network service components, the tenant SDN controller is regarded as a VNF;
- *NFV orchestrator (NFVO)*, which is responsible for central orchestration, contains a resource orchestrator (RO) and a network service orchestrator (NSO). The RO is responsible for orchestrating NFVI resources. For example, it determines the rerouted virtual paths for SFC flows, including both the VNF to NFV node remapping and the consequent subflow to virtual link remapping, as well as the processing and transmission resource scaling for the services. It also determines the virtual link to physical path (re-)mapping, to facilitate dynamic virtual link provisioning. The NSO is responsible for the life cycle management of network services, including service instantiation and dynamic network service capacity

scaling. For example, it triggers flow migration and resource scaling requests to the RO when potential QoS violations (due to traffic load fluctuations) are predicted.

1.2 Motivation and Research Topics

With SDN-NFV integration, tenants (e.g., service providers) request network services according to SLAs negotiated with the InP. The InP customizes multiple network services over a common physical infrastructure, generating service-level network slices [4, 5, 9, 16]. For each service, VNFs are embedded/placed at NFV nodes, and inter-VNF subflows are routed over physical paths between the corresponding upstream and downstream VNFs. This process is referred to as *SFC embedding*, as illustrated in Fig. 1.6, where unicast SFCs are embedded to a physical network [17–21].

At the initial network planning, SFC embedding is based on a target traffic load. The resource demands of the SFC flows are static and estimated from QoS requirements and long-term traffic statistics. Accordingly, the VNFs and subflows are allocated with static computing resources and transmission resources. With SFC embedding, a virtual path is established for each SFC flow in the virtual resource pool.

During virtual network operation, when data traffic actually enters the network, the traffic load is dynamic and can deviate from the target value. In a core network, there can be traffic dynamics in different time granularities. For example, the traffic

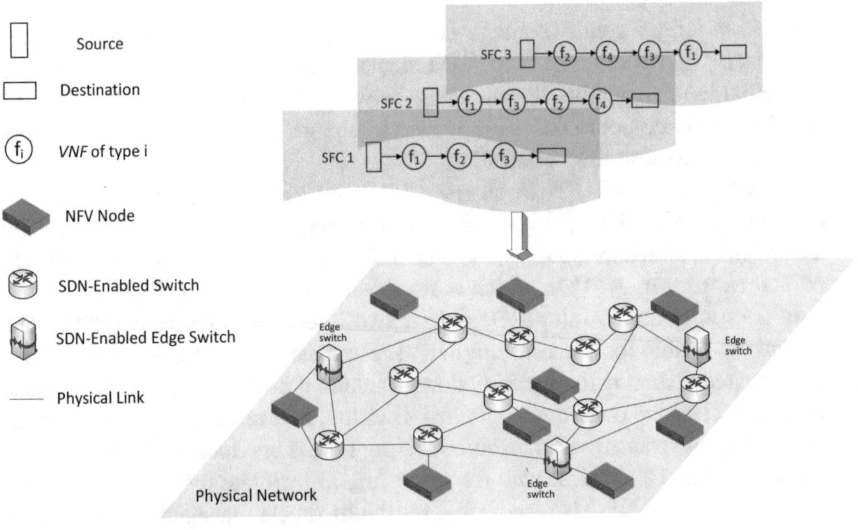

Fig. 1.6 SFC embedding to the physical network

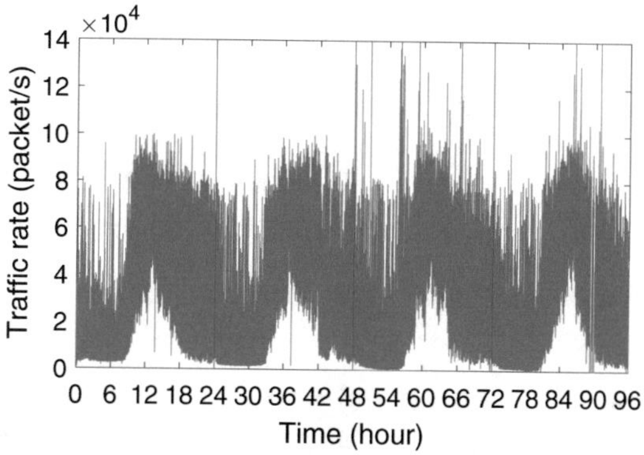

Fig. 1.7 A backbone traffic trace for the HTTP service in four days

statistics (e.g., mean and variance) can be non-stationary and experience significant changes in a coarse time granularity, which are usually predictable [22–24]. Within a long time duration with stationary traffic statistics, there are traffic dynamics in small timescales, which are usually highly bursty and unpredictable. Figure 1.7 shows a backbone traffic trace for the HTTP service in four days, which exhibits both large-timescale traffic non-stationarity with a daily periodic pattern and small-timescale traffic burstiness.

To accommodate the traffic dynamics in different time granularities, dynamic resource management among the embedded services is necessary, to ensure efficient and fair service provisioning via virtual networks with QoS guarantee. Otherwise, network congestion can occur, and services can experience performance degradation such as long queueing delay or packet loss due to delay violation. Traditional resource management methods developed for transmission-only networks cannot be directly applied in the service-oriented 5G and beyond core networks in the dual-resource context. The heterogeneity between the processing and transmission resources needs to be property dealt with.

This book is to study the dynamic resource management for the embedded services which share the processing and transmission resources in the network. We study three research topics. The first two topics address dynamic resource provisioning for the embedded services to accommodate the large-timescale traffic statistical changes, while the third topic focuses on dynamic resource scheduling for the embedded services in the presence of the small-timescale traffic dynamics.

In the first research topic, we investigate dynamic flow migration for the embedded services, to avoid QoS degradation due to the mismatch between traffic load and resource availability at the initial virtual path. With flow migration, the SFC flows can migrate to alternative virtual paths with elastic resource allocations. Since many state-dependent VNFs are associated with locally updated states for accurate processing, the states should be transferred to the new location if a VNF is migrated. The state transfers require transmission resource overhead and incur extra latency. How to achieve a trade-off between load balancing and migration overhead during each flow migration is the focus of the first research topic. In the second research

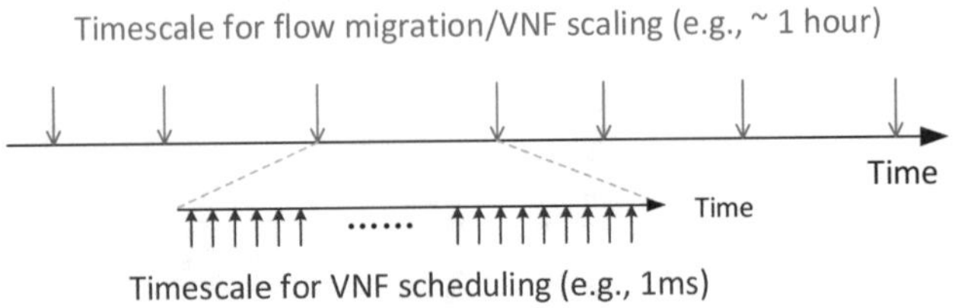

Fig. 1.8 An illustration of the timescales for dynamic resource management

topic, we further investigate two research questions in flow migration. First, when to trigger resource scaling and possible flow migrations to accommodate large-timescale traffic variations? With traffic fluctuations, a lightly loaded NFV node can become heavily loaded in the future due to increasing background traffic load, and vice versa. How to adapt to traffic patterns and achieve a trade-off between load balancing and migration cost in the long run is the other question. To answer the two questions, we focus on the VNF scaling issue in a local network segment with several candidate NFV nodes, and use a real-world non-stationary traffic trace as traffic input. Although multiple VNFs can be deployed at a common NFV node, how to schedule the central processing unit (CPU) computing resources among them to achieve efficient and fair resource sharing in the presence of small-timescale traffic dynamics should be studied. Therefore, the third research topic is on the VNF scheduling within a sufficiently long time duration, given VNF placement and stationary traffic statistics. The different timescales for flow migration (or VNF scaling) and VNF scheduling are illustrated in Fig. 1.8.

Next, we present an overview of the research topics.

1.2.1 Dynamic Flow Migration

With SFC embedding, a virtual path is established for each SFC flow in the virtual resource pool. At the initial network planning stage, SFC embedding is based on a target traffic load. During virtual network operation, when data traffic actually enters the network, the traffic load is dynamic and can deviate from the target value, possibly overloading some NFV nodes and virtual links while underloading some others from time to time. Imbalanced load can create bottlenecks on NFV nodes or virtual links, leading to QoS degradation and possible network congestion for the affected services. To avoid QoS violation caused by the load-resource mismatch, each SFC should be provisioned elastically, with dynamically provisioned computing and transmission resources. For example, an SFC flow can be dynamically migrated to an alternative virtual path with scaled computing and transmission resources, to balance traffic load in the network. Here, we consider flow migration for SFCs in a virtual resource pool, and focus on the service-level connectivity. We consider both vertical scaling and migration under the assumption that the total number of VNF instances is unchanged. Hence, the flow migration problem consists of two joint subproblems, namely, (1) finding the new VNF placement at the NFV nodes and the corresponding new subflow to virtual link remapping, and (2) processing resource scaling for the VNFs and transmission resource scaling for the subflows. Since the virtual link to which a subflow is mapped is uniquely determined by the locations of the corresponding upstream and downstream VNFs, the subflow to virtual link remapping is a byproduct of a flow migration decision. However, to maintain the infrastructure-level connectivity of SFCs, the potential topology update requirements for the virtual resource pool is considered as an overhead for flow migration. How to find physical paths for the new virtual links and how to scale up/down the existing virtual links by allocating more/less resources on the corresponding physical paths can be addressed by the typical traffic engineering (TE) methods in the transmission-only networks, such as by solving a multi-commodity flow problem.

Service flow migration, i.e., steering SFC flows through alternative NFV nodes and virtual links, is also a TE approach for elastic SFC provisioning [25]. Extensive studies have been done on TE to find paths for data delivery from source to destination within link capacity [26, 27]. A cost function, such as a piece-wise linear increasing and convex function of link utilization, can be used to penalize high link utilization near capacity. The traditional TE ensures that no packets get sent across overloaded links, by minimizing link utilization costs. Similarly, for service flow migration, the maximum loading on NFV nodes can be minimized, to achieve load balancing over processing resources.

However, traditional TE methods cannot be directly applied in service-oriented 5G and beyond networks due to the following reasons:

- Candidate paths for an SFC flow must traverse through NFV nodes for processing. In traditional TE problems, a flow is a source-destination pair without a predefined sequence of intermediate processing nodes;
- The unique properties of processing resources should be considered, such as the VNF-dependent processing density and the switching overhead at each NFV node embedded with multiple VNFs for CPU scheduling;
- For QoS provisioning to delay-sensitive services, the E2E delay requirement for each service should be considered in flow migration;
- During flow migration, the states of the migrated VNFs should be transferred to target NFV nodes for state consistency and processing inaccuracy. Hence, VNF state transfers should be taken into consideration in modeling the reconfiguration overhead for flow migration.

The state transfers lead to transmission resource overhead and incur extra latency. To minimize the migration overhead incurred by state transfers, the total number of state transfers should be minimized, which is equivalent to minimizing the modifications to the current VNF placement at the NFV nodes. In this case, the loads on different NFV nodes can be rather imbalanced, with some heavily loaded and some lightly loaded, which can possibly result in more migrations in the future. Therefore, we have another goal to minimize the maximum NFV node utilization

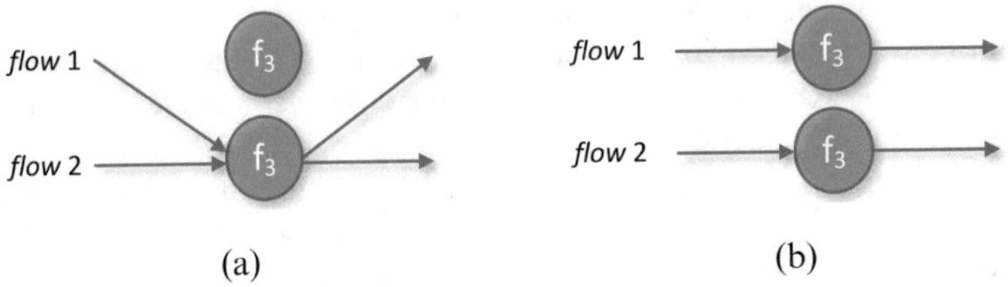

Fig. 1.9 An illustration for the trade-off between migration overhead and load balancing. (**a**) Less migration more imbalanced load. (**b**) Less imbalanced load more migration

factor to achieve load balancing. The two goals can conflict with each other. For example, a pure load balancing solution may result in frequent VNF migrations. Figure 1.9 illustrates the trade-off between load balancing and migration overhead in a simple scenario. Suppose that solution (a) is the current solution. With an increase of traffic, solution (a) works as long as the resource on the bottom NFV node is sufficient to satisfy the delay requirement. No migration takes place, but the resource utilization of the bottom NFV node is high. In solution (b), flow 1 is migrated to the top NFV node to relieve the high load on the bottom one, to have more balanced load between the two NFV nodes at the cost of one migration. With the consideration that not every two NFV nodes are directly connected by virtual links, the number of new virtual links required for flow rerouting should also be minimized to reduce the signaling overhead. Therefore, we should jointly consider the three objectives in the flow migration.

In Chap. 3, we present a delay-aware flow migration scheme for the embedded services, to guarantee the average E2E delay for each service, while addressing the trade-off between load balancing and migration overhead, under resource capacity constraints and maximal tolerable service downtime constraints. The service chaining requirements and the unique properties of processing resources such as the VNF-dependent processing density and the switching overhead are taken into consideration. A multi-objective mixed integer optimization problem is formulated, which minimizes a weighted summation of the maximum NFV node utilization factor, the total transmission resource overhead incurred by state transfers, and the number of new virtual links for flow rerouting. For delay awareness, under the assumption of Poisson traffic model and prior knowledge of time-varying packet arrival rates, the average E2E delay requirements are included in constraints based on M/M/1 queue based delay modeling. Under the assumption of sufficient transmission resources, we ignore the delay on virtual links. The computing resource constraints are incorporated with the consideration of VNF dependent processing density and switching overhead. Due to several quadratic constraints, an optimal solution to the problem is difficult to obtain using solvers such as Gurobi. We transform the original problem into a tractable mixed integer quadratically constrained programming (MIQCP) problem. Although the two problems are not equivalent, it is proved that there is a zero optimality gap between them. Given an MIQCP optimum, the optimum of the original problem is obtained through a mapping algorithm. The MIQCP transformation, together with the mapping algorithm, gives an optimal solution, but time complexity is high due to NP-hardness of the MIQCP problem. Therefore, a low-complexity heuristic algorithm based on redistribution of hop delay bounds can be used to obtain a sub-optimal solution to the original problem [28–30].

1.2.2 Dynamic VNF Resource Scaling and Migration

In Chap. 3, to focus on the trade-off between load balancing and migration cost in a multi-service scenario, the dynamic flow migration model is based on a simplified M/M/1 queueing model at the VNFs, under the assumption of Poisson traffic and exponential service time distribution for VNF packet processing. The Poisson traffic of an SFC has a predictable changing rate across non-overlapping time intervals. An implicit assumption is that the starting and ending times for each time interval are known, and the time duration of the intervals are the same. In such a traffic model, the large-timescale traffic statistical changes are represented by the Poisson traffic rate variations over known time intervals. An SFC is modeled as a tandem of M/M/1 queues at the VNFs, based on which the average E2E delay requirement for an SFC can be expressed in closed form as a constraint in the optimization problem. As demonstrated in Sect. 3.5, the flow migration optimization model guarantees the average E2E delay for Poisson traffic input, but only around 67% packets experience an E2E delay below the average. Moreover, the delay performance is degraded as traffic burstiness increases.

In reality, the core network traffic is more bursty than Poisson traffic, the delay requirement is usually more stringent than an average delay requirement, and the traffic statistical changes happen in non-regular time intervals with variable lengths. Therefore, it is necessary to remove the Poisson traffic and equal-length time interval assumptions and consider a more stringent delay requirement. Suppose that the E2E delay bound of a service is decomposed into per-hop delay bounds at each VNF. Then, a probabilistic delay requirement at a VNF requires that the probability of packet processing (including queueing) delay at a VNF exceeding a certain delay bound should be limited to an upper bound. To satisfy the probabilistic delay requirement in the presence of traffic variations, we can resort to resource demand prediction, to predict the time-varying resource demands following the traffic statistical changes. Then, VNF scaling decisions can be made, to scale up/down the amount of resources allocated to the VNFs according to the predicted resource demands, and to update the placement of VNFs among several candidate NFV nodes.

Since the real-world traffic usually exhibits non-stationary traffic characteristics across intervals with uncertain time durations, the decision epoch lengths for dynamic resource scaling and VNF migration should be adaptively adjusted to follow the changes in traffic statistics (e.g., mean and variance) and resource demands. At each decision epoch, a VNF scaling decision is made, which possibly requires

VNF migrations. To focus on traffic analysis and resource demand prediction, we consider one VNF in a local network segment with several NFV nodes, and treat the dynamics of other VNFs as background traffic at the NFV nodes. The dynamics of other VNFs are attributed to dynamics in both their traffic arrivals and their scaling decisions. To address the trade-off between load balancing and migration cost, we jointly consider the two objectives, by jointly minimizing the migration cost and the maximum resource utilization factor among all candidate NFV nodes. Moreover, there is a trade-off between cost minimizations in the short term and in the long run. When a VNF migration is required, the VNF is migrated to the current most lightly loaded NFV node for cost minimization in the current decision epoch. However, a lightly loaded NFV node can become heavily loaded in the future due to increasing background resource usage, resulting in further migrations to avoid performance degradation. In contrast, for cost minimization in the long run, the VNF should be migrated to an NFV node which is expected not to be heavily loaded in the current and successive decision epochs. Reinforcement learning (RL) provides an approach for long-run cost minimization, with the ability to capture inherent patterns in network dynamics and to make intelligent decisions accordingly [11, 23, 31–40].

In Chap. 4, we present a learning-based dynamic VNF scaling scheme, to adaptively trigger and perform VNF resource scaling and migration decisions based on detected traffic statistical changes in a real-world non-stationary traffic trace, in order to consistently satisfy the probabilistic delay requirement. A traffic parameter learning based on change point detection is presented, to learn traffic parameters of each detected stationary traffic segment in a real-world non-stationary traffic trace, based on which resource demand prediction can be performed. Then, we use a deep-Q learning approach for VNF migration decision, which learns the traffic pattern and addresses the trade-off between load balancing and migration cost in the long run. The unique features of the VNF resource scaling and migration scheme are summarized as follows, with a workflow given in Fig. 1.10 [41].

- Under the assumption that a non-stationary traffic trace can be partitioned into consecutive stationary traffic segments with unknown change points, the change points between consecutive stationary segments are identified based on change point detection. Each stationary traffic segment corresponds to one decision

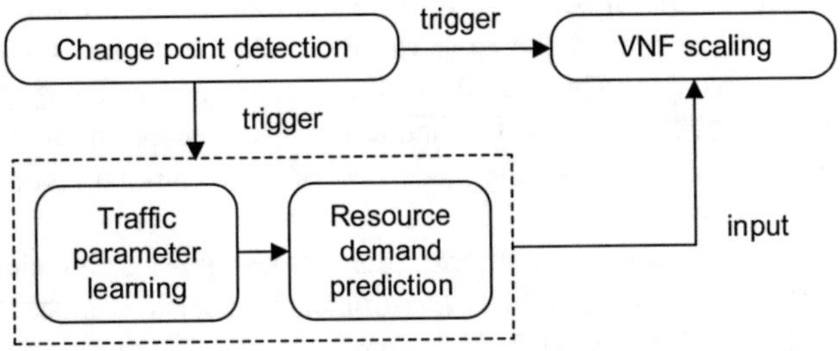

Fig. 1.10 Workflow of dynamic VNF scaling over non-stationary traffic

epoch. We use a Bayesian online change point detection (BOCPD) algorithm to detect statistical changes in mean and variance of the medium-timescale (e.g., 20 s) traffic time series. The algorithm provides online estimation of a probability distribution of current run length and the most probable mean and variance at each medium-timescale traffic sample. A run is defined as a traffic segment with the same statistics. Then, we use a threshold-based policy to identify deterministic change points (boundaries) between consecutive stationary traffic segments. The employed machine learning tool is conjugate Bayesian analysis;

- After a new change point is detected, the traffic parameters of the upcoming stationary traffic segment should be learned. Since the BOCPD algorithm is statistical, it results in a latency between the real change points and the detected change points. We exploit the latency for a *look-back* traffic parameter learning scheme. A number of small-timescale (e.g., 100 ms) traffic samples before the detected change point are collected for traffic model regression. Since the core network traffic has a high aggregation level, the Gaussian traffic approximation works well beyond a timescale of around 100 ms. Hence, we adopt the fractional Brownian motion (fBm) traffic model for each stationary traffic segment to incorporate Gaussianity and other properties of real-world core network traffic such as self-similarity and long-range dependence (LRD) [42, 43]. Then, the fBm traffic parameters can be learned through training a Gaussian process regression (GPR) model with a selected fBm covariance (kernel) function. Afterwards, the resource demand of the upcoming stationary traffic segment for a required QoS performance is calculated using empirical models;
- Change point detection provides a triggering signal for VNF scaling. Hence, the length of a VNF scaling decision epoch is varying and depends on change point detection. With the detected change points and predicted resource demands, a VNF migration problem is formulated as a Markov decision process (MDP) with variable-length decision epochs, to minimize the overall cost integrating imbalanced loading, migration cost, and resource overloading penalty in the long run. A deep Q-learning algorithm with penalty-aware prioritized experience replay is proposed to solve the MDP, with performance gains in terms of both cost and training loss reduction compared with benchmark algorithms.

1.2.3 Dynamic VNF Scheduling

The VNF placement and resource allocation should be adjusted based on traffic statistical changes in large time granularities. Within a sufficiently long time duration, e.g., an hour, with stationary traffic statistics for each service, the VNF placement at the NFV nodes remains unchanged, and the CPU computing resource budgets at the NFV nodes for the services are fixed. In a given VNF placement plan, an NFV node can hold multiple VNFs of different services with CPU computing resource sharing. For an NFV node with multiple CPU cores, there is a *multiple-to-multiple* mapping between VNFs and CPU cores, i.e., each CPU core can support multiple VNFs, and each VNF can be supported by multiple CPU cores. As each VNF corresponds to a software process, the CPU computing resource scheduling in the NFV environment is at the software process level. Once a VNF is scheduled, it occupies the associate CPUs for a certain time duration and a batch of packets are processed. Two VNFs deployed at different CPU cores can be scheduled simultaneously, but at most one VNF can be scheduled to occupy the CPU resources for packet processing at each CPU core at a given time instant.

The time granularity for VNF (software process) scheduling should not be too small, to avoid frequent switching overhead between different scheduled VNF processes, such as the CPU scheduling overhead for selecting the next process to run and the context switching time overhead for recording and loading contexts [44]. The minimum time quantum in some *state-of-the-art* operating system (OS) process schedulers is in the 100 μs to *ms* timescale, such as 100 μs for the completely fair scheduler (CFS) and 1 ms for the real time scheduler (RLS) [45]. Some OS schedulers developed for the NFV system can support a smaller time quantum such as 10 μs, but they are still in the initial development stage [46]. Hence, VNF scheduling scheme should use a realistic and *state-of-the-art* time quantum. For each CPU core, we should determine which VNF to schedule, while achieving efficient and fair resource sharing among services in the presence of small-timescale traffic dynamics.

With a *state-of-the-art* time quantum in the timescale of 100 μs to *ms*, the sequences for scheduling the VNFs of different services with CPU resource sharing have a significant impact on the QoS performance of each service. For example, if we use a time quantum (or time slot length) of 1 ms for VNF scheduling, all packets of a service with an E2E deadline of 10 ms will expire if only VNFs from other services are scheduled within 10 time slots in an extreme case. Hence, each service should be given a fair opportunity of CPU occupation to guarantee the individual QoS requirements. Due to the inter-service coupling of the VNF scheduling decisions, how to coordinate the VNF scheduling for different services with fair resource sharing and delay guarantee is complex.

The VNF scheduling decisions are also intra-SFC coupled, since the E2E delay performance of a service depends on the scheduling decisions of all VNFs in the chain. For example, if a VNF temporarily experiences congestion with a long queueing delay, the congestion can be relieved by temporarily halting the scheduling of the upstream VNF. If a VNF is not congested, the upstream VNF can be scheduled temporarily, since the packets processed from the upstream VNF will not overburden the considered VNF and create a long queueing delay there. Hence, the VNF scheduling decisions within a service should be jointly considered for the E2E delay guarantee.

In Chap. 5, we study a delay-aware VNF scheduling problem in the presence of bursty and unpredictable small-timescale traffic dynamics, to coordinate the VNF scheduling for different deadline-constrained services with delay guarantee. A delay-aware online VNF scheduling algorithm is presented, which determines the VNF to be scheduled at each NFV node during each time slot. The main features and advantages of the delay-aware VNF scheduling algorithm are summarized as follows.

- We use a packet delay aware queueing model for each service, by introducing virtual packet processing queues augmented with packet delay information at each VNF, which is the foundation for developing a delay-aware VNF scheduling algorithm;
- For efficient and fair utilization of the allocated resources at the NFV nodes, the VNF scheduling problem is formulated as a stochastic offline problem which maximizes a total network utility with proportional fairness among services, while stabilizing all the VNF packet processing queues and satisfying the delay requirements of all the services. The stochastic offline problem is transformed into an online problem by decoupling the VNF scheduling decisions over time slots with the Lyapunov optimization technique, based on which an online VNF scheduling algorithm is derived [18, 47]. Distributed VNF scheduling decisions are made at each NFV node for each time slot, based on the observed local network status such as packet arrivals and queue length.

1.3 Outline

The rest of this book is organized as follows. The three research topics, including the problems and solutions, are presented in Chaps. 3, 4, and 5, respectively.

In Chap. 3, we first present the system model under consideration, and then formulate the delay-aware flow migration problem. For tractability, we transform the original optimization problem into an MIQCP problem, and derive the optimality gap between the transformed problem and the original problem. Last, performance evaluation for both the MIQCP and heuristic solutions is presented, followed by a summary.

In Chap. 4, after presenting the system model under consideration, we introduce the change-point-driven traffic parameter learning and resource demand prediction schemes, followed by the MDP formulation and the penalty-aware deep Q-learning algorithm for dynamic VNF migration. This chapter also ends with performance evaluation and a summary.

In Chap. 5, we first present the system model under consideration, including the delay-aware virtual packet processing queueing model, and then formulate a stochastic offline delay-aware VNF scheduling problem. Based on Lyapunov optimization, an online distributed delay-aware VNF scheduling algorithm is derived, whose performance is then evaluated. A summary of this chapter is given in the end.

Finally, concluding remarks are drawn and future research directions are discussed in Chap. 6.

References

1. Vision, I.: Framework and overall objectives of the future development of IMT for 2020 and beyond. International Telecommunication Union (ITU), Document, Radiocommunication Study Groups (2015)
2. Silva, I.D., Ayoubi, S.E., Boldi, O., Bulakci, Ö., Spapis, P.: 5G RAN Architecture and Functional Design. Tech. rep. (2016)
3. Bennis, M., Debbah, M., Poor, H.V.: Ultrareliable and low-latency wireless communication: tail, risk, and scale. Proc. IEEE **106**(10), 1834–1853 (2018)
4. Zhuang, W., Ye, Q., Lyu, F., Cheng, N., Ren, J.: SDN/NFV-empowered future IoV with enhanced communication, computing, and caching. Proc. IEEE **108**(2), 274–291 (2020)
5. Ordonez-Lucena, J., Ameigeiras, P., Lopez, D., Ramos-Munoz, J.J., Lorca, J., Folgueira, J.: Network slicing for 5G with SDN/NFV: concepts, architectures, and challenges. IEEE Commun. Mag. **55**(5), 80–87 (2017)
6. Li, X., Djukic, P., Zhang, H.: Zoning for hierarchical network optimization in software defined networks. In: Proc. IEEE NOMS, pp. 1–8 (2014)
7. Zhang, N., Zhang, S., Yang, P., Alhussein, O., Zhuang, W., Shen, X.: Software defined space-air-ground integrated vehicular networks: challenges and solutions. IEEE Commun. Mag. **55**(7), 101–109 (2017)
8. Herrera, J.G., Botero, J.F.: Resource allocation in NFV: a comprehensive survey. IEEE Trans. Netw. Serv. Manag. **13**(3), 518–532 (2016)
9. Nguyen, V.G., Brunstrom, A., Grinnemo, K.J., Taheri, J.: SDN/NFV-based mobile packet core network architectures: a survey. IEEE Commun. Surv. Tutorials **19**(3), 1567–1602 (2017)
10. Shen, X., Gao, J., Wu, W., Lyu, K., Li, M., Zhuang, W., Li, X., Rao, J.: AI-assisted network-slicing based next-generation wireless networks. IEEE Open J. Veh. Technol. **1**(1), 45–66 (2020)
11. Wu, W., Chen, N., Zhou, C., Li, M., Shen, X., Zhuang, W., Li, X.: Dynamic RAN slicing for service-oriented vehicular networks via constrained learning. IEEE J. Sel. Areas Commun. **39**(7), 2076–2089 (2021)
12. Duan, Q., Ansari, N., Toy, M.: Software-defined network virtualization: an architectural framework for integrating SDN and NFV for service provisioning in future networks. IEEE Netw. **30**(5), 10–16 (2016)
13. McKeown, N., Anderson, T., Balakrishnan, H., Parulkar, G., Peterson, L., Rexford, J., Shenker, S., Turner, J.: Openflow: enabling innovation in campus networks. ACM SIGCOMM Comput. Commun. Rev. **38**(2), 69–74 (2008)
14. Lorenz, C., Hock, D., Scherer, J., Durner, R., Kellerer, W., Gebert, S., Gray, N., Zinner, T., Tran-Gia, P.: An SDN/NFV-enabled enterprise network architecture offering fine-grained security policy enforcement. IEEE Commun. Mag. **55**(3), 217–223 (2017)
15. ETSI NFV ISG: NFV-EVE005: SDN usage in NFV architectural framework (2015)
16. Ye, Q., Li, J., Qu, K., Zhuang, W., Shen, X., Li, X.: End-to-end quality of service in 5G networks—examining the effectiveness of a network slicing framework. IEEE Veh. Technol. Mag. **13**(2), 65–74 (2018)
17. Alhussein, O., Do, P.T., Ye, Q., Li, J., Shi, W., Zhuang, W., Shen, X., Li, X., Rao, J.: A virtual network customization framework for multicast services in NFV-enabled core networks. IEEE J. Sel. Areas Commun. **38**(6), 1025–1039 (2020)
18. Chen, X., Ni, W., Collings, I.B., Wang, X., Xu, S.: Automated function placement and online optimization of network functions virtualization. IEEE Trans. Commun. **67**(2), 1225–1237 (2019)
19. Xu, Z., Liang, W., Huang, M., Jia, M., Guo, S., Galis, A.: Efficient NFV-enabled multicasting in SDNs. IEEE Trans. Commun. **67**(3), 2052–2070 (2019)
20. Li, D., Hong, P., Xue, K., et al.: Virtual network function placement considering resource optimization and SFC requests in cloud datacenter. IEEE Trans. Parallel Distrib. Syst. **29**(7), 1664–1677 (2018)

21. Ye, Q., Zhuang, W., Li, X., Rao, J.: End-to-end delay modeling for embedded VNF chains in 5G core networks. IEEE Internet Things J. **6**(1), 692–704 (2019)
22. Fei, X., Liu, F., Xu, H., Jin, H.: Adaptive VNF scaling and flow routing with proactive demand prediction. In: Proc. IEEE INFOCOM, pp. 486–494 (2018)
23. Luo, Z., Wu, C., Li, Z., Zhou, W.: Scaling geo-distributed network function chains: A prediction and learning framework. IEEE J. Sel. Areas Commun. **37**(8), 1838–1850 (2019)
24. MAWI Working Group Traffic Archive (2020). http://mawi.wide.ad.jp/mawi/. Accessed 17 Dec 2020
25. Tang, H., Zhou, D., Chen, D.: Dynamic network function instance scaling based on traffic forecasting and VNF placement in operator data centers. IEEE Trans. Parallel Distrib. Syst. **30**(3), 530–543 (2019)
26. Fortz, B., Thorup, M.: Internet traffic engineering by optimizing OSPF weights. In: Proc. IEEE INFOCOM, pp. 519–528 (2000)
27. Rexford, J.: Route optimization in IP networks. Handbook of Optimization in Telecommunications pp. 679–700 (2006)
28. Qu, K., Zhuang, W., Ye, Q., Shen, X., Li, X., Rao, J.: Dynamic flow migration for embedded services in SDN/NFV-enabled 5G core networks. IEEE Trans. Commun. **68**(4), 2394–2408 (2020)
29. Qu, K., Zhuang, W., Ye, Q., Shen, X., Li, X., Rao, J.: Traffic engineering for service-oriented 5G networks with SDN-NFV integration. IEEE Netw. **34**(4), 234–241 (2020)
30. Qu, K., Zhuang, W., Ye, Q., Shen, X., Li, X., Rao, J.: Delay-aware flow migration for embedded services in 5G core networks. In: Proc. IEEE ICC, pp. 1–6 (2019)
31. Sutton, R.S., Barto, A.G.: Reinforcement Learning: An Introduction. MIT Press, Cambridge (2011)
32. Chinchali, S., Hu, P., Chu, T., Sharma, M., Bansal, M., Misra, R., Pavone, M., Katti, S.: Cellular network traffic scheduling with deep reinforcement learning. In: Proc. AAAI'18 (2018)
33. Wang, J., Zhao, L., Liu, J., Kato, N.: Smart resource allocation for mobile edge computing: a deep reinforcement learning approach. IEEE Trans. Emerg. Topics Comput. **9**(3), 1529–1541. https://doi.org/10.1109/TETC.2019.2902661
34. Li, H., Ota, K., Dong, M.: Learning IoT in edge: deep learning for the Internet of things with edge computing. IEEE Netw. **32**(1), 96–101 (2018)
35. Li, J., Shi, W., Zhang, N., Shen, X.: Delay-aware VNF scheduling: a reinforcement learning approach with variable action set. IEEE Trans. Cogn. Commun. Netw. **7**(1), 304–318 (2020). https://doi.org/10.1109/TCCN.2020.2988908
36. Liu, J., Guo, H., Xiong, J., Kato, N., Zhang, J., Zhang, Y.: Smart and resilient EV charging in SDN-enhanced vehicular edge computing networks. IEEE J. Sel. Areas Commun. **38**(1), 217–228 (2020)
37. Wu, W., Cheng, N., Zhang, N., Yang, P., Zhuang, W., Shen, X.: Fast mmwave beam alignment via correlated bandit learning. IEEE Trans. Wirel. Commun. **18**(12), 5894–5908 (2019)
38. Ye, Q., Shi, W., Qu, K., He, H., Zhuang, W., Shen, X.: Joint RAN slicing and computation offloading for autonomous vehicular networks: a learning-assisted hierarchical approach. IEEE Open J. Veh. Technol. **2**, 272–288 (2021). https://doi.org/10.1109/OJVT.2021.3089083
39. Wu, W., Yang, P., Zhang, W., Zhou, C., Shen, X.: Accuracy-guaranteed collaborative DNN inference in industrial IoT via deep reinforcement learning. IEEE Trans. Ind. Inform. **17**(7), 4988–4998 (2021)
40. Zhou, C., Wu, W., He, H., Yang, P., Lyu Feng and, C.N., Shen, X.: Deep reinforcement learning for delay-oriented IoT task scheduling in space-air-ground integrated network. IEEE Trans. Wirel. Commun. **20**(2), 911–925 (2021)
41. Qu, K., Zhuang, W., Ye, Q., Shen, X., Li, X., Rao, J.: Dynamic resource scaling for VNF over nonstationary traffic: a learning approach. IEEE Trans. Cogn. Commun. Netw. **7**(2), 648–662 (2021)
42. Cheng, Y., Zhuang, W., Wang, L.: Calculation of loss probability in a finite size partitioned buffer for quantitative assured service. IEEE Trans. Commun. **55**(9), 1757–1771 (2007)

43. Fraleigh, C., Tobagi, F., Diot, C.: Provisioning IP backbone networks to support latency sensitive traffic. In: Proc. IEEE INFOCOM, pp. 1871–1879 (2003)
44. Emmerich, P., Raumer, D., Gallenmüller, S., Wohlfart, F., Carle, G.: Throughput and latency of virtual switching with open vswitch: a quantitative analysis. J. Netw. Syst. Manag. **26**(2), 314–338 (2018)
45. Kulkarni, S.G., Zhang, W., Hwang, J., Rajagopalan, S., Ramakrishnan, K., Wood, T., et al.: NFVnice: dynamic backpressure and scheduling for NFV service chains. IEEE/ACM Trans. Netw. **28**(2), 639–652 (2020)
46. Chowdhury, S.R., Bai, T., Boutaba, R., François, J., et al.: UNiS: a user-space non-intrusive workflow-aware virtual network function scheduler. In: 2018 14th International Conf. on Network and Service Management (CNSM), pp. 152–160 (2018)
47. Mao, Y., Zhang, J., Song, S., Letaief, K.B.: Stochastic joint radio and computational resource management for multi-user mobile-edge computing systems. IEEE Trans. Wirel. Commun. **16**(9), 5994–6009 (2017)

Chapter 2
Literature Review

2.1 Service Function Chain (SFC) Embedding

We refer to the SFC embedding for a given set of service requests with static resource demands as static SFC embedding. The static SFC embedding problem has been extensively studied for unicast services [1–10]. The orchestration of SFCs poses two correlated subproblems, i.e., how to place the VNFs at NFV nodes, and how to route the inter-VNF subflows in the physical network. For each subflow, it must be routed over physical paths between two NFV nodes where the upstream and downstream VNFs of the subflow are placed. The joint problem of VNF placement and traffic routing should consider computing and transmission resource consumption. Typically, the VNFs are placed to a minimal feasible number of NFV nodes to reduce the VNF deployment cost, while the inter-VNF subflows are routed over feasible physical paths between the corresponding upstream and downstream VNFs with minimal transmission resource consumption. The joint problem is usually modeled as a mixed integer linear programming (MILP) problem to minimize the total computing and transmission resource provisioning cost, under the computing resource capacity constraints at the NFV nodes and the transmission resource capacity constraints at the physical links. There are also some studies on static SFC embedding for multicast services such as video conferencing, multi-player augmented reality games, and file distribution [11–16]. The multicast routing topology is jointly determined by the locations of VNFs and multicast replication points. In [15, 16], joint VNF placement and multicast traffic routing is considered, in which a single multicast replication point is assumed in the multicast routing topology. All the destinations of a multicast service share the same set of VNF instances, and the multicast replication point occurs after all the VNFs in the multicast routing topology. This brings inefficient transmission resource utilization especially when the destinations are geographically dispersed. Alhussein *et al.* develop a flexible multicast routing and NF placement framework for both single-service and multi-service scenarios, by introducing multiple multicast replication

W. Zhuang, K. Qu, *Dynamic Resource Management in Service-Oriented Core Networks*, Wireless Networks, https://doi.org/10.1007/978-3-030-87136-9_2

points and moving them earlier in the multicast routing topology, and allowing duplicated VNF deployment just after the multicast replication points [17]. Then, the traffic flow for each destination can traverse through a different set of VNF instances, providing flexibility for multicast topology customization. The VNF deployment cost and link provisioning cost are jointly minimized in designing the multicast routing topology.

2.2 Elastic SFC Provisioning

There are works on dynamic SFC embedding with both service request dynamics and traffic dynamics [18–21]. With the dynamic arrivals of new services requests, the VNFs and subflows of the new requests should be embedded to the physical network, while the VNF placement and traffic routing of the existing services are dynamically adjusted. To accommodate the dynamic traffic of the existing services, the SFC embedding results should also be dynamically adjusted. In this section, we provide an overview of existing works on elastic SFC provisioning from the aspects of reconfiguration overhead awareness, QoS awareness, triggering scheme, and data-driven adaptivity.

2.2.1 Reconfiguration Overhead Awareness

Existing studies on dynamic SFC embedding take into account the trade-off between resource consumption cost and operational overhead due to reconfiguration [18–20]. The reconfiguration overhead is usually modeled as a weighted number of reconfigured NFV nodes and physical links [19], or the total revenue loss due to throughput loss within a constant service downtime [20], or the time duration for all state transfers associated with flow migration [22]. In [19], the migration scheme of VNFs in dynamic SFCs is studied to minimize the costs from resource consumption and reconfiguration. Genetic algorithms are proposed to handle the VNF migration and SFC routing in response to the change of user workload. Given the new demands, the VNFs are re-deployed on servers with re-allocated server resources, and the traffic between consecutive VNFs are rerouted on physical paths with re-allocated transmission resources on the paths. The reconfiguration cost is modeled as a linear combination of the number of reconfigured servers and the number of reconfigured links, based on which the trade-off between resource consumption cost and reconfiguration cost is studied. However, the fundamental reasons why such reconfiguration cost should be minimized need to be clearly understood.

Recent studies have revealed that one of the main problems in migrating traffic to alternative NFV nodes is the fact that many network functions are state-dependent [23]. Some examples of state-dependent VNFs include a virtual intrusion detection system (IDS) keeping track of pattern matchings for accurate attack detection in a stream of packets, and a virtual network address translator (NAT) storing mappings between ports and hosts. For state-dependent VNFs, the packet header or payload processing relies on VNF states, and the states are consistently updated together with packet header or payload processing, to guarantee accurate processing for subsequent packets in a traffic flow. The VNF states are stored and updated locally in associate VNFs. It introduces state inconsistency if the in-progress traffic flows on a state-dependent VNF are rerouted to an alternative NFV node without transferring the associate VNF states to the target NFV node, which brings packet processing inaccuracy for the subsequent packets and hinders seamless service migration. Hence, during flow migration, the states of the migrated VNFs should be transferred to target NFV nodes for consistency.

Some frameworks such as OpenNF are proposed to solve the state inconsistency problem [23]. At a state-dependent VNF where flow migration occurs, not only the incoming packets of the migrated flow to the VNF are rerouted to the target NFV node, but also the associate VNF states are transferred to the target NFV node [23–25]. During a VNF state transfer, packets are halted until the states in the original NFV node are completely transferred to the target NFV node, which incurs extra latency to the affected packets. To reduce the overall state transfer latency for an SFC with multiple state-dependent VNFs in migration, Co-Scalar [26] provides a parallel state transfer scheme which allows states of multiple VNFs to be transferred in parallel, thus greatly reducing the latency overhead at a cost of transmission resource overhead for simultaneous state transfers.

Hence, VNF state transfers should be taken into consideration in modeling the reconfiguration overhead for flow migration. One performance metric for flow migration is the maximum allowable service downtime within a certain time duration [27]. Under the assumption that the time interval for flow migration is much larger than state transfer time, we consider to minimize the total transmission resource overhead incurred by state transfers within a maximal tolerable service downtime in each service interruption due to flow migration.

2.2.2 *QoS Awareness*

For delay-sensitive services in 5G and beyond core networks, the resource demand is dependent on both the statistics of traffic arrivals and the delay requirement. With changes in traffic statistics, the computing resource demand of the VNFs and the transmission resource demand of the inter-VNF subflows both vary to satisfy a given delay requirement.

Existing studies on dynamic SFC embedding usually assume prior knowledge about the time-varying resource demands or predict the future resource demands based on historical resource demand information, based on which VNFs are placed at alternative NFV nodes, and inter-VNF subflows are rerouted over alternative physical paths [18–20, 28–30]. The QoS requirements are expressed in such a way that the time-varying resource demands should be satisfied without exceeding the resource capacity, but the delay requirement is not explicitly addressed. In [28, 30], the average traffic rate in a certain time duration is taken as the average resource demand in packet/s, and dynamic SFC embedding is performed based on the changing average traffic rate. However, resource allocation/scaling according to the average traffic rate is not sufficient to satisfy a stringent delay requirement in the presence of bursty traffic.

In [31–33], both the resource capacity and delay constraints are taken into consideration for dynamic SFC provisioning, but with a focus on the link propagation delay in the E2E delay modeling. The VNF packet processing delay and load dependent queueing delay are ignored. In [33], a delay violation penalty is characterized as a function of link propagation delays over the E2E path for each SFC flow, which indicates the cost if the E2E delay requirement of an SFC is violated. A delay violation cost minimization problem is then formulated to determine traffic routing for each SFC flow. In [32], the E2E delay including packet processing delay, packet queueing delay, and packet transmission delay is considered in multicast virtual network embedding. The delay is determined using network traffic measuring tools instead of analytical modeling.

For embedded SFCs, how CPU and bandwidth resources are shared among multiple SFCs determines the share of processing and transmission rates available for each SFC flow, and thus affects E2E packet delay calculation. Some analytical frameworks for average E2E delay modeling have been proposed for embedded SFCs, based on Poisson traffic assumption. With such a traffic model, the large-timescale traffic statistical changes are represented by changes in the Poisson traffic rate. Under the assumption of exponential packet processing delay, an analytical E2E delay model for packets passing through an embedded SFC is established in [34], where an M/M/1/K queueing model is employed to evaluate packet delay at each VNF, based on which the average E2E delay requirement for an SFC can be expressed in closed form. Similarly, an M/M/1 queueing model is used in [35] to characterize the packet delay at each VNF. In [36], a tandem queueing model is developed to characterize the E2E delay of an embedded SFC, with an M/D/1 queueing model for the first VNF in each chain, and approximated M/D/1 queueing models for the subsequent VNFs in the chain.

However, the real-world traffic in core networks shows multi-timescale traffic dynamics, including predictable large-timescale traffic statistical changes, and non-predictable traffic burstiness in small time granularity [37]. For delay-sensitive services, the delay requirement is usually expressed as a more strict probabilistic constraint rather than an average delay requirement. For example, the probability of VNF packet processing delay exceeding a certain delay bound should not exceed an upper limit. This probability is usually small, such as 0.01, 0.001, and even smaller values. In the presence of real-world traffic dynamics, it is difficult to guarantee such a stringent probabilistic delay requirement, based on Poisson traffic model.

Moreover, a real-world resource demand trace with inherent delay guarantee is difficult to obtain. Instead, a traffic trace with packet arrival information is usually available [38]. Therefore, a resource demand prediction scheme is required, to predict the time-varying QoS-aware resource demands following the traffic statistical changes in an available packet arrival traffic trace.

2.2.3 *Resource Scaling Triggers*

To enable elastic SFC provisioning, a triggering scheme is required, to determine when to perform resource scaling and possible VNF migration. In existing studies on dynamic VNF placement and traffic routing, time is usually partitioned into consecutive decision epochs of equal length, e.g., 30 min, and dynamic decisions are made in a proactive or reactive manner at each decision instant [20, 28, 30]. The selection of epoch length is difficult and usually based on experience. If the decision epoch is too long, traffic burstiness in different time granularities within an epoch cannot be captured, resulting in challenges for continuous QoS guarantee; if the decision epoch is too short, decisions are made frequently, possibly resulting in unnecessary expensive VNF migrations for temporal short traffic bursts.

A better way is to adopt adaptive epoch length according to changes in traffic statistics (e.g., mean and variance) and resource demands, since the real-world traffic usually exhibits non-stationary traffic characteristics across intervals with uncertain time durations. Several change point detection algorithms, either retrospective or online, have been developed for detecting structural breaks in a non-stationary time series [39–41]. Online algorithms provide inference about change points as each data sample arrives, which is more appropriate for detecting statistical changes in a non-stationary traffic time series, based on which VNF scaling decisions can be made reactively without a significant latency [40, 41]. Under the assumption that a non-stationary traffic trace can be partitioned into consecutive stationary traffic segments with unknown change points, the decision epochs with variable lengths are to be identified based on change point detection. Each stationary traffic segment corresponds to one decision epoch.

2.2.4 *Data-Driven Adaptivity*

Due to the recent advances in deep learning and deep reinforcement learning (RL), the literature is enjoying a renewed interest in the application of (deep) RL on the dynamic SFC provisioning problem [35, 42–44]. RL provides an approach for long-run cost minimization, with the ability to capture inherent patterns in network dynamics and to make intelligent decisions accordingly [28, 45–50]. Pei et al. study the dynamic VNF placement (or VNF migration) problem in an NFV-enabled network under a dynamic traffic trace with periodic patterns [35], by using a combination of optimization and reinforcement learning approaches. The network substrate is divided into smaller regions, and a double deep Q-learning algorithm is used to intelligently select network regions where the VNF placement needs adjustment. With the selected network regions, an optimization model based VNF placement algorithm is used to generate the new VNF placement results in the regions. Troia et al. consider an NFV-enabled metro-core optical network represented as a multi-layer network to maximize the number of successfully routed VNF chains, while minimizing the reconfiguration penalty, blocking probability and power consumption [42]. Gu et al. investigate dynamic VNF orchestration and flow scheduling for NFV-enabled network services with dynamically changing data rates [43]. A deep RL approach is used to relieve the unrealistic traffic model assumptions. A model-assisted RL framework is proposed, in which a heuristic algorithm guides the initial exploration during the training process, with the goal of speeding up the RL convergence. In [51], an RL approach, called parameterized action twin (PAT) deterministic policy gradient, is used to learn the elastic VNF provisioning in a cloud-RAN, including vertically scaling the VNFs with resource reconfiguration, horizontally scaling the VNFs by deploying new VNF instances, and migrating VNFs from edge to the cloud, in the presence of dynamics in network conditions, resource availability, and VNF requests from users.

2.3 Dynamic VNF Scheduling

There exist studies on delay-aware VNF scheduling with multiple services, which focus on task-level delay performance instead of packet-level delay performance [49, 52–55]. Each SFC is associated with a task, which processes a batch of packets with a given traffic block size through all the VNFs in the chain. Hence, the task of an SFC is decomposed into subtasks at each VNF in the SFC. Once a VNF is scheduled, the associate subtask is executed by using CPU computing resources at the embedded NFV node. The VNF processing time for the subtask depends on the traffic block size, the VNF type, and the maximum processing rate at the corresponding NFV node. Until a VNF finishes the processing of its subtask, its downstream VNF is allowed to be scheduled and start the processing of a succeeding subtask. Under the assumption that at most one VNF can be scheduled at each NFV node, the task completion time of each SFC depends on the VNF scheduling sequences. The task-level delay performance is optimized, by minimizing the maximum task completion time (often referred to as the makespan) of a group of SFCs, while satisfying the task completion delay requirement for each SFC.

An inherent assumption in this line of existing works is that the processing for a traffic block at different VNFs in a chain has no time overlapping, implying that the first packet in the traffic block has to wait until the last packet finishes processing, which is inefficient for services with a strict per-packet E2E deadline especially if the traffic block size is big. Hence, the algorithms developed for task-level VNF scheduling cannot be reused for deadline-constrained services with an upper limit requirement for the delay violation ratio. Also, we cannot treat a packet as a "tiny task" and reuse the existing task-level VNF scheduling algorithms for

VNF scheduling on packet-level, due to the non-negligible switching overhead between scheduling different VNF software processes. With a *state-of-the-art* time quantum, the VNF scheduling should be at least on packet batch level, which means that a batch of packets are processed at the scheduled VNF during one time slot. A challenging issue is how to guarantee the packet-level delay performance with packet batch level VNF scheduling.

Another challenging issue is how to schedule the VNFs with bursty and unpredictable traffic dynamics. With the unavailability of small-timescale traffic statistics, a potential VNF scheduling approach is an adaptation of the classical backpressure algorithm which was originally developed for transmission resources and is throughput-optimal for delay-insensitive flows [56]. With the potential approach, the differential backlogs in number of required CPU cycles is used as the VNF scheduling weight [57–59]. At each NFV node, the VNF with the current largest VNF scheduling weight is scheduled in each time slot. However, for deadline-constrained services with an upper limited delay violation ratio, such a simple adaptation without delay awareness cannot be directly applied.

2.4 Summary

In this chapter, existing studies on dynamic resource management in service-oriented 5G and beyond core networks are briefly reviewed. First, the static SFC embedding for both unicast and multicast services with static resource demands is reviewed, including the assumptions, methods, and objectives. Then, existing works on elastic SFC provisioning are summarized from the aspects of reconfiguration overhead awareness, QoS awareness, resource scaling triggers, and data-driven adaptivity. Different models for characterizing the reconfiguration overhead in dynamic SFC embedding are introduced, with a focus on the reconfiguration overhead incurred by state transfers associated with state-dependent VNFs in migration. We also provide an overview of different QoS requirements and QoS models in dynamic SFC embedding. The typical triggers for dynamic resource scaling are also briefly reviewed, including periodic triggers and change point driven triggers. The existing studies using data-driven approaches for elastic SFC provisioning based on machine learning are also briefly reviewed. Last, the existing works in task-level dynamic VNF scheduling are reviewed, and the challenges in providing packet-level delay guarantee in dynamic VNF scheduling are introduced.

References

1. Cohen, R., Lewin-Eytan, L., Naor, J.S., Raz, D.: Near optimal placement of virtual network functions. In: Proc. IEEE INFOCOM, pp. 1346–1354. IEEE, Piscataway (2015)

2. Mehraghdam, S., Keller, M., Karl, H.: Specifying and placing chains of virtual network functions. In: Proc. IEEE CloudNet, pp. 7–13 (2014)
3. Bari, M.F., Chowdhury, S.R., Ahmed, R., Boutaba, R.: On orchestrating virtual network functions. In: Proc. IEEE CNSM, pp. 50–56 (2015)
4. Luizelli, M.C., Bays, L.R., Buriol, L.S., Barcellos, M.P., Gaspary, L.P.: Piecing together the NFV provisioning puzzle: efficient placement and chaining of virtual network functions. In: Proc. IFIP/IEEE IM, pp. 98–106 (2015).
5. Ghaznavi, M., Shahriar, N., Kamali, S., Ahmed, R., Boutaba, R.: Distributed service function chaining. IEEE J. Sel. Areas Commun. **35**(11), 2479–2489 (2017)
6. Xie, Y., Liu, Z., Wang, S., Wang, Y.: Service function chaining resource allocation: a survey (2016). Accessed 14 July 2021
7. Zhang, S.Q., Zhang, Q., Tizghadam, A., Park, B., Bannazadeh, H., Boutaba, R., Leon-Garcia, A.: Sector: TCAM space aware routing on SDN. In: Int. Teletraffic Congr., pp. 216–224 (2016)
8. Zhang, N., Yang, P., Zhang, S., Chen, D., Zhuang, W., Liang, B., Shen, X.S.: Software defined networking enabled wireless network virtualization: challenges and solutions. IEEE Netw. **31**(5), 42–49 (2017)
9. Li, J., Shi, W., Ye, Q., Zhuang, W., Shen, X., Li, X.: Online joint VNF chain composition and embedding for 5g networks. In: Proc. IEEE Globecom, pp. 1–6. IEEE (2018)
10. Ye, Q., Li, J., Qu, K., Zhuang, W., Shen, X., Li, X.: End-to-end quality of service in 5G networks—examining the effectiveness of a network slicing framework. IEEE Veh. Technol. Mag. **13**(2), 65–74 (2018)
11. Zeng, M., Fang, W., Zhu, Z.: Orchestrating tree-type VNF forwarding graphs in inter-DC elastic optical networks. J. Lightw. Technol. **34**(14), 3330–3341 (2016)
12. Xu, Z., Liang, W., Huang, M., Jia, M., Guo, S., Galis, A.: Approximation and online algorithms for NFV-enabled multicasting in SDNs. In: Proc. IEEE ICDCS, pp. 625–634 (2017)
13. Kuo, J.J., Shen, S.H., Yang, M.H., Yang, D.N., Tsai, M.J., Chen, W.T.: Service overlay forest embedding for software-defined cloud networks. In: Proc. IEEE ICDCS, pp. 720–730 (2017)
14. Alhussein, O., Do, P.T., Li, J., Ye, Q., Shi, W., Zhuang, W., Shen, X., Li, X., Rao, J.: Joint VNF placement and multicast traffic routing in 5G core networks. In: Proc. IEEE Globecom, pp. 1–6 (2018)
15. Zhang, S.Q., Zhang, Q., Bannazadeh, H., Leon-Garcia, A.: Routing algorithms for network function virtualization enabled multicast topology on SDN. IEEE Trans. Netw. Serv. Manag. **12**(4), 580–594 (2015)
16. Zhang, S.Q., Tizghadam, A., Park, B., Bannazadeh, H., Leon-Garcia, A.: Joint NFV placement and routing for multicast service on SDN. In: Proc. IEEE NOMS, pp. 333–341 (2016)
17. Alhussein, O., Do, P.T., Ye, Q., Li, J., Shi, W., Zhuang, W., Shen, X., Li, X., Rao, J.: A virtual network customization framework for multicast services in NFV-enabled core networks. IEEE J. Sel. Areas Commun. **38**(6), 1025–1039 (2020)
18. Liu, J., Lu, W., Zhou, F., Lu, P., Zhu, Z.: On dynamic service function chain deployment and readjustment. IEEE Trans. Netw. Serv. Manag. **14**(3), 543–553 (2017)
19. Rankothge, W., Le, F., Russo, A., Lobo, J.: Optimizing resource allocation for virtualized network functions in a cloud center using genetic algorithms. IEEE Trans. Netw. Service Manag. **14**(2), 343–356 (2017)
20. Eramo, V., Miucci, E., Ammar, M., Lavacca, F.G.: An approach for service function chain routing and virtual function network instance migration in network function virtualization architectures. IEEE/ACM Trans. Netw. **25**(4), 2008–2025 (2017)
21. Alhussein, O., Zhuang, W.: Robust online composition, routing and NF placement for NFV-enabled services. IEEE J. Sel. Areas Commun. **38**(6), 1089–1101 (2020)
22. Xia, J., Pang, D., Cai, Z., Xu, M., Hu, G.: Reasonably migrating virtual machine in NFV-featured networks. In: Proc. IEEE Conf. Computer and Information Technology, pp. 361–366 (2016)
23. Gember-Jacobson, A., Viswanathan, R., Prakash, C., Grandl, R., Khalid, J., Das, S., Akella, A.: OpenNF: enabling innovation in network function control. In: Proc. ACM SIGCOMM, pp. 163–174 (2014)

24. Peuster, M., Karl, H.: E-state: Distributed state management in elastic network function deployments. In: Proc. NetSoft, pp. 6–10 (2016)
25. Nobach, L., Rimac, I., Hilt, V., Hausheer, D.: Statelet-based efficient and seamless NFV state transfer. IEEE Trans. Netw. Serv. Manag. **14**(4), 964–977 (2017)
26. Zhang, B., Zhang, P., Zhao, Y., Wang, Y., Luo, X., Jin, Y.: Co-scaler: cooperative scaling of software-defined NFV service function chain. In: Proc. IEEE Conf. Network Function Virtualization and Software Defined Networks, pp. 33–38 (2016)
27. Zhang, F., Liu, G., Fu, X., Yahyapour, R.: A survey on virtual machine migration: challenges, techniques, and open issues. IEEE Commun. Surveys Tuts. **20**(2), 1206–1243 (2018)
28. Luo, Z., Wu, C., Li, Z., Zhou, W.: Scaling geo-distributed network function chains: a prediction and learning framework. IEEE J. Sel. Areas Commun. **37**(8), 1838–1850 (2019)
29. Dräxler, S., Karl, H., Mann, Z.Á.: JASPER: joint optimization of scaling, placement, and routing of virtual network services. IEEE Trans. Netw. Serv. Manag. **15**(3), 946–960 (2018)
30. Fei, X., Liu, F., Xu, H., Jin, H.: Adaptive VNF scaling and flow routing with proactive demand prediction. In: Proc. IEEE INFOCOM, pp. 486–494 (2018)
31. Guo, L., Pang, J., Walid, A.: Dynamic service function chaining in SDN-enabled networks with middleboxes. In: Proc. IEEE ICNP, pp. 1–10 (2016)
32. Ayoubi, S., Assi, C., Shaban, K., Narayanan, L.: Minted: multicast virtual network embedding in cloud data centers with delay constraints. IEEE Trans. Commun. **63**(4), 1291–1305 (2015)
33. Bari, F., Chowdhury, S.R., Ahmed, R., Boutaba, R., Duarte, O.C.M.B.: Orchestrating virtualized network functions. IEEE Trans. Netw. Serv. Manag. **13**(4), 725–739 (2016)
34. Chua, F.C., Ward, J., Zhang, Y., Sharma, P., Huberman, B.A.: Stringer: balancing latency and resource usage in service function chain provisioning. IEEE Internet Comput. **20**(6), 22–31 (2016)
35. Pei, J., Hong, P., Pan, M., Liu, J., Zhou, J.: Optimal VNF placement via deep reinforcement learning in SDN/NFV-enabled networks. IEEE J. Sel. Areas Commun. **38**(2), 263–278 (2020)
36. Ye, Q., Zhuang, W., Li, X., Rao, J.: End-to-end delay modeling for embedded VNF chains in 5G core networks. IEEE Internet Things J. **6**(1), 692–704 (2019)
37. Paxson, V., Floyd, S.: Wide area traffic: the failure of Poisson modeling. IEEE/ACM Trans. Netw. **3**(3), 226–244 (1995)
38. MAWI Working Group Traffic Archive (2021). http://mawi.wide.ad.jp/mawi/. Accessed 14 July 2021
39. Liu, S., Yamada, M., Collier, N., Sugiyama, M.: Change-point detection in time-series data by relative density-ratio estimation. Neural Netw. **43**, 72–83 (2013)
40. Adams, R.P., MacKay, D.J.: Bayesian online changepoint detection. Tech. rep., University of Cambridge, Cambridge, UK (2007)
41. Comert, G., Bezuglov, A.: An online change-point-based model for traffic parameter prediction. IEEE Trans. Intell. Transp. Syst. **14**(3), 1360–1369 (2013)
42. Troia, S., Alvizu, R., Maier, G.: Reinforcement learning for service function chain reconfiguration in NFV-SDN metro-core optical networks. IEEE Access **7**, 167944–167957 (2019)
43. Gu, L., Zeng, D., Li, W., Guo, S., Zomaya, A.Y., Jin, H.: Intelligent VNF orchestration and flow scheduling via model-assisted deep reinforcement learning. IEEE J. Sel. Areas Commun. **38**(2), 279–291 (2020)
44. Fu, X., Yu, F.R., Wang, J., Qi, Q., Liao, J.: Service function chain embedding for NFV-enabled IoT based on deep reinforcement learning. IEEE Commun. Mag. **57**(11), 102–108 (2020)
45. Sutton, R.S., Barto, A.G.: Reinforcement Learning: an Introduction. MIT Press, Cambridge (2011)
46. Chinchali, S., Hu, P., Chu, T., Sharma, M., Bansal, M., Misra, R., Pavone, M., Katti, S.: Cellular network traffic scheduling with deep reinforcement learning. In: Proc. AAAI (2018)
47. Wang, J., Zhao, L., Liu, J., Kato, N.: Smart resource allocation for mobile edge computing: a deep reinforcement learning approach. IEEE Trans. Emerg. Topics Comput. **9**(3), 1529–1541 (2019). https://doi.org/10.1109/TETC.2019.2902661
48. Li, H., Ota, K., Dong, M.: Learning IoT in edge: deep learning for the Internet of things with edge computing. IEEE Netw. **32**(1), 96–101 (2018)

49. Li, J., Shi, W., Zhang, N., Shen, X.: Delay-aware VNF scheduling: a reinforcement learning approach with variable action set. IEEE Trans. Cogn. Commun. Netw. **7**(1), 304–318 (2020)
50. Liu, J., Guo, H., Xiong, J., Kato, N., Zhang, J., Zhang, Y.: Smart and resilient EV charging in SDN-enhanced vehicular edge computing networks. IEEE J. Sel. Areas Commun. **38**(1), 217–228 (2020)
51. Roig, J.S.P., Gutierrez-Estevez, D.M., Gündüz, D.: Management and orchestration of virtual network functions via deep reinforcement learning. IEEE J. Sel. Areas Commun. **38**(2), 304–317 (2020)
52. Qu, L., Assi, C., Shaban, K.: Delay-aware scheduling and resource optimization with network function virtualization. IEEE Trans. Commun. **64**(9), 3746–3758 (2016)
53. Yang, S., Li, F., Yahyapour, R., Fu, X.: Delay-sensitive and availability-aware virtual network function scheduling for NFV. IEEE Trans. Serv. Comput. https://doi.org/10.1109/TSC.2019.2927339
54. Zhang, Y., He, F., Sato, T., Oki, E.: Network service scheduling with resource sharing and preemption. IEEE Trans. Netw. Service Manag. **17**(2), 764–778 (2020)
55. Alameddine, H., Tushar, M.H.K., Assi, C.: Scheduling of low latency services in softwarized networks. IEEE Trans. Cloud Comput. **9**(3), 1220–1235 https://doi.org/10.1109/TCC.2019.2907949
56. Bui, L., Srikant, R., Stolyar, A.: Novel architectures and algorithms for delay reduction in back-pressure scheduling and routing. In: Proc. IEEE INFOCOM, pp. 2936–2940 (2009)
57. Feng, H., Llorca, J., Tulino, A.M., Molisch, A.F.: Optimal dynamic cloud network control. IEEE/ACM Trans. Netw. **26**(5), 2118–2131 (2018)
58. Gu, L., Zeng, D., Tao, S., Guo, S., Jin, H., Zomaya, A.Y., Zhuang, W.: Fairness-aware dynamic rate control and flow scheduling for network utility maximization in network service chain. IEEE J. Sel. Areas Commun. **37**(5), 1059–1071 (2019)
59. Chen, X., Ni, W., Collings, I.B., Wang, X., Xu, S.: Automated function placement and online optimization of network functions virtualization. IEEE Trans. Commun. **67**(2), 1225–1237 (2019)

Chapter 3
Dynamic Flow Migration: A Model-Based Optimization Approach

3.1 System Model

A time-slotted system is considered, in which time is partitioned into equal-length time intervals indexed by k. The VNF placement remains unchanged during one time interval, but may change across different time intervals. A flow migration decision is made for each time interval.

3.1.1 *Services*

Let $\mathcal{R}$ denote the set of embedded services. A service, $r \in \mathcal{R}$, is represented in the form of SFC. It originates from source node $\mathrm{n}_1^{(r)}$ and traverses through H_r VNFs in sequence towards destination node $\mathrm{n}_2^{(r)}$, with average E2E delay requirement D_r, maximal tolerable service downtime Ω_r in one service interruption, average packet size b_r in bit, and time-varying traffic rate $\lambda^{(r)}(k)$ in packet/s. Under the assumption that the time interval is much larger than $\max_{r\in\mathcal{R}} \Omega_r$, the experienced service downtime is much shorter than stable service operation time for any service. Let $\mathcal{H}_r = \{1, \cdots, H_r\}$, and denote the h-th ($h \in \mathcal{H}_r$) VNF in SFC r as $V_h^{(r)}$. Let $V_0^{(r)}$ and $V_{H_r+1}^{(r)}$ be dummy VNFs in SFC r, locating at source node $\mathrm{n}_1^{(r)}$ and destination node $\mathrm{n}_2^{(r)}$ respectively. Let $\mathcal{V}$ be a set containing all VNFs belonging to different SFCs, with $(r, h) \in \mathcal{V}$ denoting the h-th VNF in SFC r. Let $\mathcal{A}$ be a set of edge switches hosting all dummy VNFs. The h-th ($h \in \{0\} \cup \mathcal{H}_r$) inter-VNF subflow in SFC r, i.e., the subflow between upstream (dummy) VNF $V_h^{(r)}$ and downstream (dummy) VNF $V_{h+1}^{(r)}$, is denoted as $Y_h^{(r)}$.

W. Zhuang, K. Qu, *Dynamic Resource Management in Service-Oriented Core Networks*, Wireless Networks, https://doi.org/10.1007/978-3-030-87136-9_3

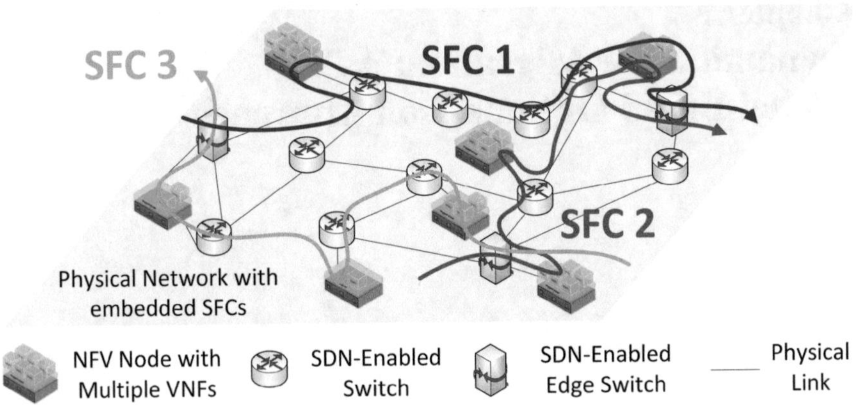

Fig. 3.1 A physical network with embedded SFCs

3.1.2 Virtual Resource Pool

Figure 3.1 shows a physical network with three embedded SFCs in single-path routing.

A virtual resource pool is abstracted from the physical network with embedded SFCs, represented as a directed graph $G = \{\mathcal{N} \cup \mathcal{A}, \mathcal{E}\}$, where $\mathcal{N}$ is a set of all NFV nodes, $\mathcal{A}$ is a set of edge switches hosting dummy VNFs, and $\mathcal{E}$ is a set of virtual links. Here, we assume that each NFV node can hold all types of VNFs for simplicity. In reality, the supported VNF types at each NFV node can be differentiated, and the placement for a VNF of a certain type should be restricted to the NFV nodes which support the same type. For virtual link $e \in \mathcal{E}$, we use binary parameters, $\{I_n^{e,1}\}$ and $\{I_n^{e,2}\}$, to describe its location and direction, with $I_n^{e,1} = 1$ if $n \in \mathcal{N} \cup \mathcal{A}$ is its starting point and $I_n^{e,2} = 1$ if $n \in \mathcal{N} \cup \mathcal{A}$ is its ending point. It is possible that G is not fully connected. Assume that there are sufficient transmission resources in the physical network. We can increase resources on existing virtual links, and find paths with enough resources for new virtual links. Hence, we consider a computing resource limited virtual resource pool.

3.1.3 Computing Model

In this subsection, we discuss the computing model within an NFV node deployed with multiple VNFs from different SFCs. We first model each VNF processing system as an M/M/1 queue, with which the relationship between the traffic rate, packet processing rate, and processing delay at a VNF can be characterized. Then, we introduce a processing density model for a VNF, which captures the relationship between the packet processing rate (in packet/s) and CPU computing resources (in cycle/s). Finally, we describe how the CPU computing resources at an NFV node are shared among multiple VNFs.

3.1.3.1 M/M/1 Queueing Model

Packet arrivals of an SFC during a sufficiently large time interval are modeled as a Poisson process, with different rates (in packet/s) across different time intervals. Specifically, traffic arrival of SFC $r \in \mathcal{R}$ during time interval k is modeled as Poisson with rate $\lambda^{(r)}(k)$ in packet/s.

Let $S^{rh}(k)$ denote the packet processing rate in packet/s at VNF $V_h^{(r)}$ during interval k, and assume that the packet processing time at each VNF follows an exponential distribution [1, 2]. Let random variable $t^{rh}(k)$ (> 0) denote the packet processing time at VNF $V_h^{(r)}$ during interval k. The probability distribution function of $t^{rh}(k)$ is given by $f(t^{rh}(k)) = S^{rh}(k)e^{-S^{rh}(k)t^{rh}(k)}$.

Then, the packet processing system at a VNF is modeled as an M/M/1 queue, with different arrival and service rates across different time intervals. Specifically, the M/M/1 queue at VNF $V_h^{(r)}$ during interval k has arrival rate $\lambda^{(r)}(k)$ and service rate $S^{rh}(k)$, both in packet/s. To guarantee the stability of M/M/1 queue, the arrival and service rates should satisfy $\lambda^{(r)}(k) \leq S^{rh}(k)$. Under the queue stability constraint, the average queueing delay at VNF $V_h^{(r)}$ during interval k, denoted by $d^{rh}(k)$, can be calculated based on the Little's Law, given by $d^{rh}(k) = \frac{1}{S^{rh}(k)-\lambda^{(r)}(k)}$.

Since an SFC is a concatenation of VNF processing systems, an SFC can be modeled as a linear M/M/1 queueing network, as illustrated in Fig. 3.2. Then, the average E2E processing delay of SFC $r \in \mathcal{R}$ during interval k is represented as

$$\sum_{h \in \mathcal{H}_r} d^{rh}(k) = \sum_{h \in \mathcal{H}_r} \left(\frac{1}{S^{rh}(k) - \lambda^{(r)}(k)} \right). \tag{3.1}$$

With traffic load fluctuations (i.e., the variations in $\lambda^{(r)}(k)$ across different time intervals), the processing rate at each VNF should be scaled up/down via resource scaling to guarantee at least an average E2E processing delay for an SFC flow. Suppose that the E2E processing delay requirement was initially decomposed into per-hop delay requirements at each VNF. Let $\mathsf{D}^{rh}(k)$ denote the hop delay bound at

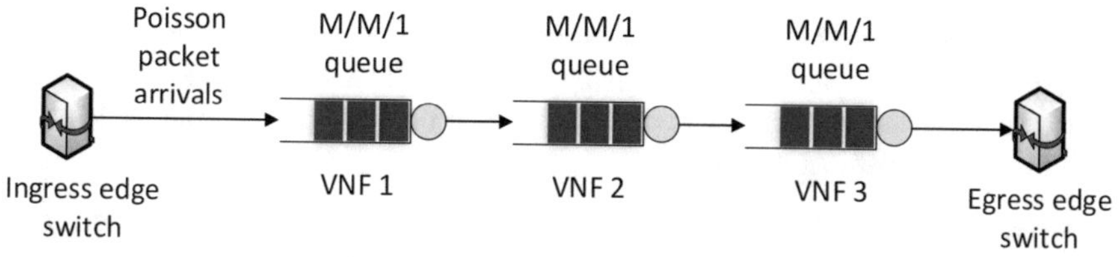

Fig. 3.2 A linear M/M/1 queueing network for an SFC with three VNFs

VNF $V_h^{(r)}$ during interval k, with $\sum_{h \in \mathcal{H}_r} \mathsf{D}^{rh}(k) = \mathsf{D}_r$ for SFC $r \in \mathcal{R}$. Based on the M/M/1 queueing model, the processing rate of VNF $V_h^{(r)}$ during interval k, $S^{rh}(k)$, should satisfy $S^{rh}(k) \geq \lambda^{(r)}(k) + \frac{1}{\mathsf{D}^{rh}(k)}$, to guarantee that the average queueing delay at VNF $V_h^{(r)}$ during interval k, $d^{rh}(k)$, is upper bounded by $\mathsf{D}^{rh}(k)$.

3.1.3.2 Processing Density Model

Before our discussion of the processing density model which characterizes the relationship between packet processing rate and CPU computing resources, we first review the relationship between packet transmission rate and transmission resources on a virtual link. The transmission resources on a virtual link are represented in bit/s. For each unit packet/s of transmission rate, the transmission resource demand in bit/s depends only on the packet size. For example, to support a transmission rate of S_1 packet/s to flow 1 with packet size b_1 bits, the transmission resource demand of flow 1 is $S_1 b_1$ in bit/s at all its traversing virtual links.

With the consideration of heterogeneity between computing and transmission resources, we should characterize the relationship between packet processing rate and CPU computing resources at an NFV node. The computing resource capacity C_n of NFV node $n \in \mathcal{N}$ is its maximum supporting CPU processing rate in cycle/s. For one packet/s of processing rate, the CPU resource demand at a certain NFV node depends on many factors, including the packet size, the type of function, the packet I/O scheme, and the virtualization technology [3–5]. Typically, packet I/O, virtualization and function processing constitute the total CPU usage at an NFV node for packet processing, which are summarized into VNF-independent and VNF-dependent parts:

- VNF-independent CPU computing resource usage: With advanced packet I/O [4, 6] and virtualization technology [5], packet copying overheads in the packet I/O subsystem and the virtualization hypervisor are eliminated, thus greatly reducing the per-bit CPU overhead. In other words, with the same packet I/O and virtualization technology across different NFV nodes, for each packet/s in processing rate, the CPU computing resource demand in cycle/s for packet I/O and virtualization at different NFV nodes is almost the same. Here, we assume it a constant;
- VNF-dependent CPU computing resource usage: In terms of function processing, some functions (such as NAT) process only packet headers, other functions such as encryption/decryption process both packet headers and payloads. Consequently, for unit packet/s in processing rate, the CPU resource demand in cycle/s for function processing at a certain NFV node depends on both the packet size and the type of function.

Considering the two parts of CPU usage, the CPU computing resource demand of a VNF at an NFV node consists of a constant part and a VNF-dependent part.

Define processing density of VNF $V_h^{(r)}$ as $P_h^{(r)}$ (in cycle/bit), which is the CPU computing resource demand (in cycle/s) of VNF $V_h^{(r)}$ for one bit/s of processing rate. Accordingly, $P_h^{(r)} b_r$ is the processing density of VNF $V_h^{(r)}$ in cycle/packet. The processing density is also referred to as computing intensity.

With the queueing and processing density models, the relationships among the traffic rate, packet processing rate, computing resources, and VNF processing delay can be established.

3.1.3.3 Multi-VNF Computing Resource Sharing

Consider *multiple-to-one* mapping between VNFs and NFV nodes. For example, consider an NFV node with two incoming virtual links and two outgoing virtual links as shown in Fig. 3.3. Four SFC flows share the NFV node and its surrounding virtual links. Packets incoming from different virtual links are split into flow-specific processing queues served by the CPU. After CPU processing, packets are sent to flow-specific transmission queues associated with the corresponding virtual link. In this scenario, CPU computing resources at the NFV node are shared among all the four flows, while transmission resources at each virtual link are shared among the traversing flows.

Traditionally, transmission resources are shared among multiple flows according to the general processor sharing (GPS) discipline. For example, to support a transmission rate of S_1 packet/s to flow 1 with packet size b_1 bits and a transmission rate of S_3 packet/s to flow 3 with packet size b_3 bits over the upper outgoing virtual link in Fig. 3.3, a total amount of $S_1 b_1 + S_3 b_3$ transmission resources (in bit/s) is required, which should not exceed the maximum supporting transmission rate of the virtual link. Similarly, under the assumption of infinitely divisible resources, the GPS discipline can be employed for CPU computing resource sharing among

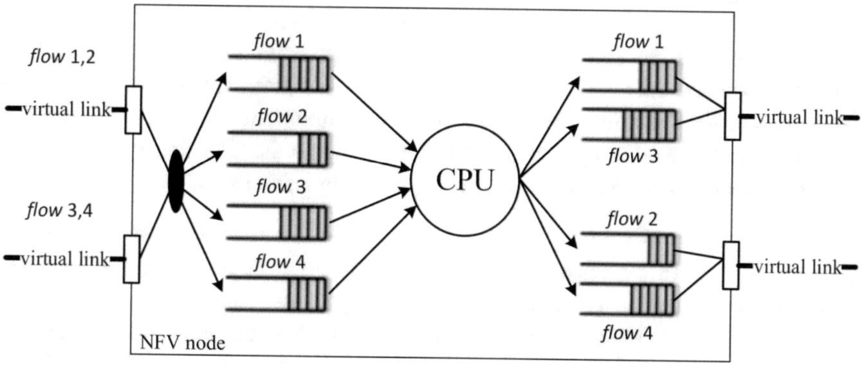

Fig. 3.3 An illustration of processing and transmission resource sharing

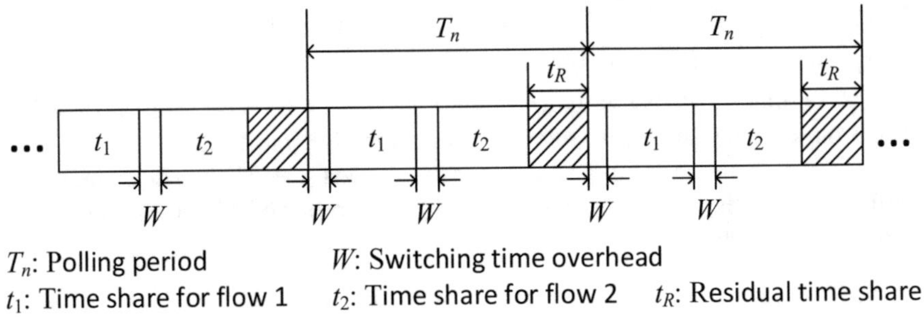

Fig. 3.4 An illustration of CPU polling period with two flows

multiple flows. The difference is the non-negligible switching overhead for CPU computing resource sharing.

Remark 3.1 Under the assumption that resources are infinitely divisible, GPS is a benchmark resource allocation scheme to achieve QoS isolation and multiplexing gain among flows [7, 8]. Without such an assumption for CPU computing resources, we assume that there exists a practical VNF scheduling algorithm (to be studied in Chap. 5) which achieves performance comparable with GPS.

Specifically, a CPU polling scheme is employed for resource sharing among multiple VNFs at an NFV node. For example, as illustrated in Fig. 3.4, two flows are polled for service periodically at NFV node $n \in \mathcal{N}$, with a period of T_n. Each flow corresponds to one VNF, and each VNF corresponds to one VNF packet processing queue. Once polled, the associated VNF occupies the CPU of the NFV node for an allocated CPU time share in a polling period. During the allocated CPU time share, only packets from the polled VNF processing queue are processed. Each flow gets a portion of CPU computing resources which is linear with its allocated CPU time share. The polling scheme introduces CPU scheduling overhead and multi-task context switching overhead, due to extra CPU time spent on determining the next VNF process to run and on saving and loading contexts between every two consecutive tasks [9]. Here, processing packets from a certain flow (or VNF processing queue) is a task. The total time overhead for switching between VNFs are referred to as the switching time overhead. The polling period T_n and the switching time overhead $\mathtt{W}$ in NFV node n are constant. Based on the GPS discipline, the two VNFs are guaranteed minimum processing rates (in packet/s) of $S_1 = \frac{t_1 \mathsf{C}_n}{T_n P_n^1 b_1}$ and $S_2 = \frac{t_2 \mathsf{C}_n}{T_n P_n^2 b_2}$ respectively, where $t_1 + t_2 + 2\mathtt{W} = T_n - t_R$, and t_R is the residual time share in a polling period. Define utilization factor of NFV node n, denoted by η_n, as the percentage of allocated time shares plus switching time overhead in a polling period of NFV node n. The switching time overhead at NFV node n linearly increases with the number of VNFs placed at the NFV node if the number of VNFs is larger than 1.

3.1.4 *Elastic SFC Provisioning*

In this subsection, we introduce elastic SFC provisioning from three aspects. First, with joint VNF migration and resource scaling, the VNFs of the SFCs can be flexibly migrated among the NFV nodes in the virtual resource pool, with elastic computing resources allocated to each VNF. Second, with flexible virtual link provisioning, the inter-VNF subflows of the SFCs can be flexibly re-mapped to alternative virtual links in the virtual resource pool, while at the same time the topology of the virtual resource pool is consistently updated. Third, to guarantee seamless and accurate processing for the migrated state-dependent VNFs, VNF state transfer schemes and the corresponding overhead are discussed.

3.1.4.1 Joint VNF Migration and Vertical Scaling

The NFV enables elastic scaling of computing resources allocated to VNFs in a cost-effective manner, which facilitates agile service provisioning and management. Dynamic VNF operations, including horizontal scaling, vertical scaling, and migration are widely employed to provide elastic VNF provisioning [10]. With horizontal scaling, the number of instances for a VNF is scaled in or out, with constant amount of computing resources for each instance. With vertical scaling, the amount of computing resources for a VNF instance is scaled up/down. With VNF migration, a VNF instance migrates to an alternative NFV node, without changing the amount of computing resources.

Here, we assume that each VNF is instantiated only once and do not consider horizontal scaling. To satisfy the time-varying computing resource demands of VNFs, the amount of computing resources allocated to a VNF at an NFV node can be vertically scaled up/down. However, VNF vertical scaling is limited by the amount of computing resources available at the NFV node. To avoid overloading and achieve load balancing, we consider joint VNF migration and vertical scaling, which means that a VNF can migrate to an alternative NFV node with sufficient resources to satisfy its computing resource scaling demand. To avoid confusion with vertical delay scaling to be introduced in Sect. 3.4.2, we refer to VNF vertical scaling as resource scaling.

Figure 3.5 illustrates joint VNF migration and resource scaling for two SFCs, with each SFC composed of two VNFs. We use rectangle height to represent the amount of computing resources. From Fig. 3.5a, we observe that both NFV nodes A and C are lightly loaded while NFV node B is heavily loaded. With significant traffic increase for both SFCs, NFV node B will be overloaded if all the VNFs remain at the current NFV nodes. In this case, the E2E delay requirements of both SFCs will be violated. To ensure the QoS performance for both SFCs, Fig. 3.5b shows a joint VNF migration and resource scaling plan after resource demand increase. Specifically, one VNF of SFC 1 migrates from NFV node B to NFV node C, and all

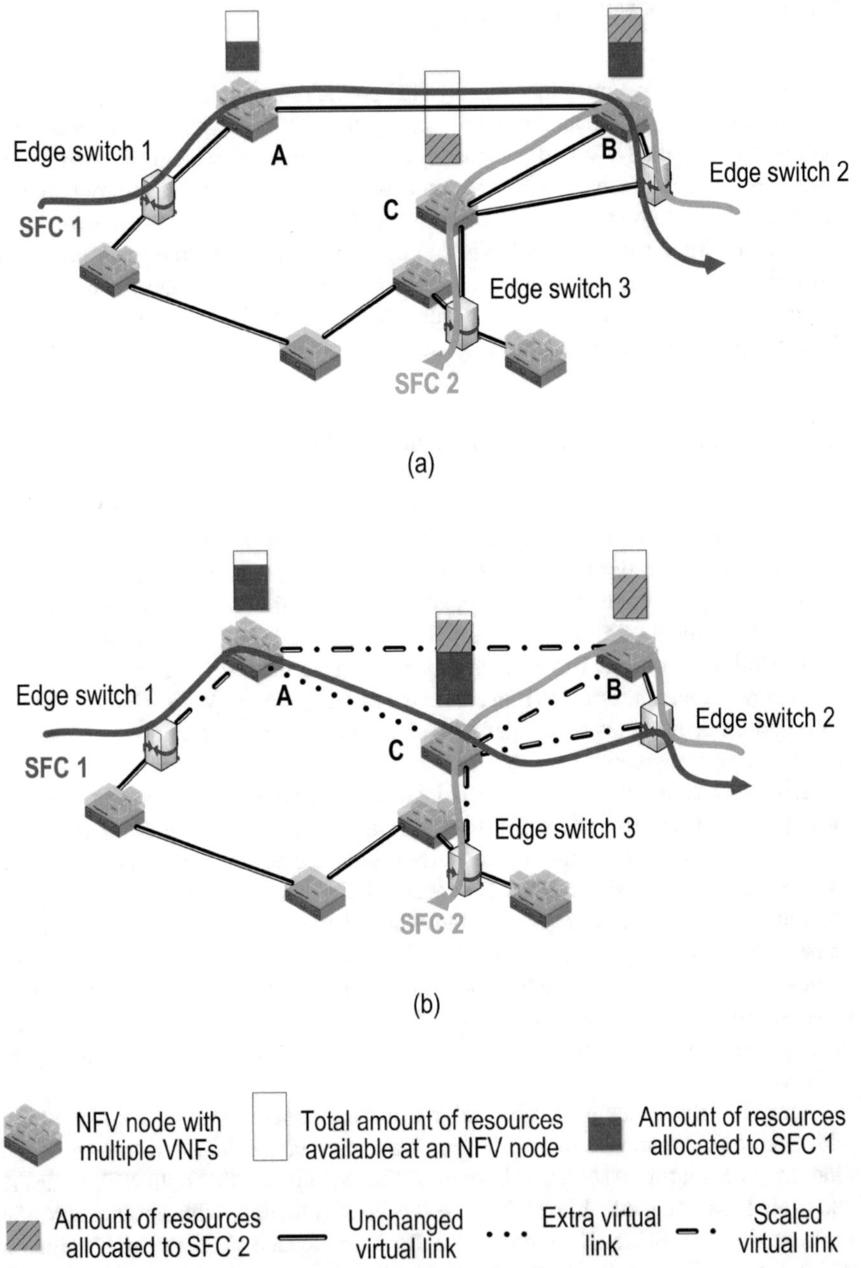

Fig. 3.5 An illustration for flow migration in a virtual resource pool with topology update. (**a**) before resource demand increase. (**b**) after resource demand increase

VNFs are vertically scaled. In this way, none of the three NFV nodes are overloaded, and QoS violation is avoided.

The joint VNF migration and resource scaling decision corresponds to two groups of decision variables. The first group represents VNF migration, and the second group corresponds to resource scaling, which are defined as follows:

- Let $\mathbf{x}(k) = \{\mathrm{x}_n^{rh}(k)\}$ be a binary decision variable set for interval k, with $\mathrm{x}_n^{rh}(k) = 1$ if (dummy) VNF $V_h^{(r)}$ is mapped to node $n \in \mathcal{N} \cup \mathcal{A}$ during interval k, and $\mathrm{x}_n^{rh}(k) = 0$ otherwise. From interval $(k-1)$ to interval k, the decision variable set representing VNF to NFV node mapping changes from $\mathbf{x}(k-1) = \{\mathrm{x}_n^{rh}(k-1)\}$ to $\mathbf{x}(k) = \{\mathrm{x}_n^{rh}(k)\}$, where $\mathbf{x}(k-1)$ is known at the beginning of interval k. We introduce binary decision variable set $\boldsymbol{g}(k) = \{g_{n,n'}^{rh}(k)\}$ for interval k, with $g_{n,n'}^{rh}(k) = 1$ if the mapped NFV node for VNF $V_h^{(r)}$ changes from NFV node $n \in \mathcal{N}$ during interval $(k-1)$ to NFV node $n' \in \mathcal{N}$ during interval k, and $g_{n,n'}^{rh}(k) = 0$ otherwise. The decision variable set $\boldsymbol{g}(k) = \{g_{n,n'}^{rh}(k)\}$ is uniquely determined by $\mathbf{x}(k) = \{\mathrm{x}_n^{rh}(k)\}$, given the values of $\mathbf{x}(k-1) = \{\mathrm{x}_n^{rh}(k-1)\}$;
- In terms of resource scaling, the packet processing rate (in packet/s) for VNF $V_h^{(r)}$ can be scaled up/down during interval k. Let $\boldsymbol{S}(k) = \{S_n^{rh}(k)\}$ be a nonnegative continuous decision variable set for interval k, with $S_n^{rh}(k)$ being the processing rate in packet/s allocated to VNF $V_h^{(r)}$ by NFV node $n \in \mathcal{N}$ during interval k.

Then, we have $S^{rh}(k) = \sum_{n \in \mathcal{N}} \mathrm{x}_n^{rh}(k) S_n^{rh}(k)$ for the packet processing rate of VNF $V_h^{(r)}$ during interval k. According to the M/M/1 queueing model in Sect. 3.1.3.1, the E2E processing delay of SFC $r \in \mathcal{R}$ can be derived, given the traffic rates $\lambda^{(r)}(k)$ and the decision variable sets $\mathbf{x}(k) = \{\mathrm{x}_n^{rh}(k)\}$ and $\boldsymbol{S}(k) = \{S_n^{rh}(k)\}$. More details are given in Sect. 3.2.

3.1.4.2 Flexible Virtual Link Provisioning

Assume that there are sufficient transmission resources and no queueing on virtual links. For a subflow, if its upstream or downstream VNF migrates to an alternative NFV node, it should be re-mapped to an alternative virtual link between the re-mapped upstream NFV node (or ingress edge switch) and downstream NFV node (or egress edge switch).

However, it is possible that the virtual resource pool $G = \{\mathcal{N} \cup \mathcal{A}, \mathcal{E}\}$ is not fully connected, i.e., there is no direct virtual link between two NFV nodes or between edge switch and NFV node. In this case, an new virtual link is required to support the re-mapped subflow between the new pair of upstream and downstream nodes. Thus, depending on the current topology of the virtual resource pool, the new virtual link for the re-mapped subflow can be an existing one or an extra one. In the former case, the transmission rate of the existing virtual link should be scaled up to support the re-mapped subflow. In the latter case, the transmission rate of the new virtual link should be at least the transmission resource demand of the re-mapped subflow.

Assume that the infrastructure SDN controller can find physical paths with enough transmission resources for a scaled-up or an new virtual link.

After a successful re-mapping, transmission resources allocated to the subflow should be released from the old virtual link. Then, the transmission rate of the old virtual link can be scaled down if there are other flows traversing it. Otherwise, the old virtual link can be removed from the virtual resource pool.

As shown in Fig. 3.5, a subflow of SFC 1 is released from virtual link $A \rightarrow B$, and re-mapped to an new virtual link $A \rightarrow C$ between NFV nodes A and C. Accordingly, the transmission rate over the existing virtual link $A \rightarrow B$ can be scaled down, under the assumption that there are remaining background traffic over virtual link $A \rightarrow B$. Without such an assumption, virtual link $A \rightarrow B$ should be deleted from the virtual resource pool. For the virtual links corresponding to the subflows with unchanged mapping, such as virtual link $B \rightarrow C$, the transmission rates should be scaled up, to support the traffic increase of traversing SFCs. With the scaling, creation, and removal of virtual links in the virtual resource pool, the virtual links are flexibly provisioned, and the topology of the virtual resource pool is consistently updated, which addresses the potential mismatch between computing resources at fixed NFV nodes and transmission resources on existing virtual links.

Let $\mathbf{y}(k) = \{y_{nn'}^{rh}(k)\}$ be a binary variable set for interval k, with $y_{nn'}^{rh}(k) = 1$ if subflow $Y_h^{(r)}$ is mapped to an new virtual link between $n, n' \in \mathcal{N} \cup \mathcal{A}$ during interval k, and $y_{nn'}^{rh}(k) = 0$ otherwise. The new virtual links for flow rerouting incur signaling overhead between the infrastructure SDN controller and SDN switches, due to forwarding rule reconfiguration along the underlying physical paths. Assume that the signaling overhead for flow rerouting over the new virtual links linearly increases with the total number of new virtual links, which is calculated as $\sum_{r\in\mathcal{R}}\sum_{h\in\{0\}\cup\mathcal{H}_r}\sum_{n,n'\in\mathcal{N}\cup\mathcal{A}} y_{nn'}^{rh}(k)$. To reduce migration cost, the total number of new virtual links should be minimized.

3.1.4.3 VNF State Transfer

Consider both state-dependent and stateless VNFs for the SFCs. When an SFC flow migrates at a state-dependent VNF, the VNF is re-mapped to an alternative NFV node, with the associate states transferred to the target NFV node. Figure 3.6 illustrates the flow migration and associate state transfers, where two VNFs are re-mapped, and two state transfers are triggered correspondingly. Packet processing is halted during state transfer, incurring a service downtime.

Let $U_h^{(r)}(k)$ (in bit) be the time-varying state size of VNF $V_h^{(r)}$, whose value at time interval k is monitored by the SDN controller. For a state transfer at VNF $V_h^{(r)}$, $U_h^{(r)}(k)$ is the product of state transfer time and consumed transmission resources (in bit/s) [11]. Without loss of generality, we consider that a stateless VNF has a "state" with state size equal zero. Then, if an SFC flow migrates at a stateless VNF, the corresponding "state transfer" incurs no state transfer overhead in terms of both state transfer time and transmission resource overhead.

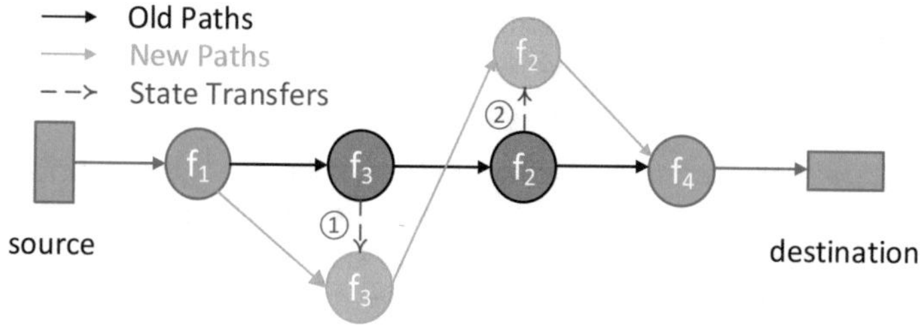

Fig. 3.6 An illustration of flow migration and state transfer

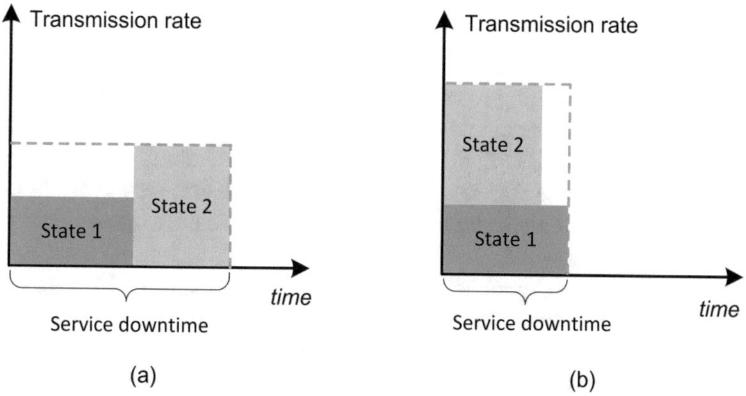

Fig. 3.7 An illustration of parallel and sequential state transfer overhead. (**a**) Sequential state transfer. (**b**) Parallel state transfer

For an SFC, if multiple VNFs are re-mapped to alternative NFV nodes, there are the same number of state transfers as the number of VNF re-mappings (migrations). For such an SFC with multiple state transfers, there are two state transfer schemes[1] with different overhead:

- If the states are transferred sequentially, the state transfer scheme is referred to as sequential state transfer. Since packet processing is halted during state transfer, the SFC experiences a service downtime equal to the aggregated state transfer time of all the state transfers, as illustrated in Fig. 3.7a. However, since the state transfers incur transmission resource consumption sequentially, the overall transmission rate overhead during the service downtime is the maximum transmission rate among all state transfers;
- If the states are transferred in parallel, the state transfer scheme is referred to as parallel state transfer [12]. As illustrated in Fig. 3.7b, the service downtime is the maximum state transfer time among all the state transfers. However, since

[1] Here we assume that a state transfer consumes the same transmission rate and requires the same state transfer time in both state transfer schemes.

the state transfers incur transmission resource consumption simultaneously, the overall transmission rate overhead during the service downtime is the aggregated transmission rate of all state transfers.

Under the assumption that a state transfer incurs the same state transfer overhead in either sequential or parallel state transfer, the service downtime for an SFC with multiple state transfers under the parallel state transfer scheme is much less than that of sequential state transfer, at the cost of transmission rate overhead. Since SFC $r \in \mathcal{R}$ has a maximal tolerable service downtime Ω_r in one service interruption, we consider parallel state transfer in which all state transfers can take place simultaneously. Under the assumption that a flow migration time interval is sufficiently long, the service downtime is much shorter than the stable service operation time for any service.

A state transfer belonging to one service can share a virtual link with subflows from other services. For example, as shown in Fig. 3.5, a state transfer from SFC 1 and a subflow from SFC 2 both happen on virtual link $B \to C$. Generally, since services experience different downtimes, transmission resources allocated to subflows can be opportunistically used by state transfers before the subflows resume packet forwarding. However, for the example in Fig. 3.5, transmission resources allocated to the subflow cannot be used by the state transfer, since SFC 2 experiences no service downtime. Thus, the transmission rate over virtual link $B \to C$ should be scaled up. In the worst case, all state transfers happen on virtual links where no subflows are mapped or no transmission resources allocated to subflows can be opportunistically used. In this case, dedicated virtual links should be established for state transfers, or the transmission resource capacity of existing virtual links should be scaled up to support state transfers.

Therefore, we consider the total amount of transmission resources required by the parallel state transfers as a overhead in flow migration. Define a nonnegative continuous variable set $\boldsymbol{B}(k) = \{B^{rh}_{n,n'}(k)\}$, with $B^{rh}_{n,n'}(k)$ denoting the transmission resource overhead to transfer the state of VNF $V^{(r)}_h$ from NFV node $n \in \mathcal{N}$ to NFV node $n' \in \mathcal{N}$ during interval k. Here, we assume that the transmission resource overhead for each state transfer is no less than $B_{min} = \frac{min(U^{(r)}_h(k))}{max(\Omega_r)}$, i.e., $B^{rh}_{n,n'}(k) \geq B_{min}$. Then, the total normalized transmission resource overhead for parallel state transfer is represented as $\sum_{(r,h)\in\mathcal{V}} \sum_{n,n'\in\mathcal{N}} \frac{B^{rh}_{n,n'}(k)}{B_{min}}$. Within a maximal tolerable service downtime, the total normalized transmission resource overhead incurred by state transfers should be minimized.

3.2 Optimization Model for Dynamic Flow Migration

Consider a set $\mathcal{R}$ of SFCs in a computing resource limited virtual resource pool. Assume sufficient transmission resources for the flexible topology update of the virtual resource pool. During time interval k, traffic arrival of SFC $r \in \mathcal{R}$ is Poisson with rate $\lambda^{(r)}(k)$ packet/s. The rate can be predicted at the end of time interval $(k-1)$ based on measurements and historical information [13, 14]. With traffic variations of the SFCs from interval $k-1$ to interval k, i.e., from $\{\lambda^{(r)}(k-1)\}$ to $\{\lambda^{(r)}(k)\}$, we consider dynamic flow migration for the SFCs, with joint VNF migration and resource scaling, which (1) determines the re-mapping between VNFs and NFV nodes, represented by binary decision variable set $\mathbf{x}(k) = \{\mathrm{x}_n^{rh}(k)\}$ as defined in Sect. 3.1.4.1,and (2) vertically scale the amount of computing resources allocated to VNFs, represented by continuous decision variable set, $\boldsymbol{S}(k) = \{S_n^{rh}(k)\}$, as defined in Sect. 3.1.4.1.

In this section, we first discuss about constraints in flow migration in terms of VNF migration, resource scaling, subflow to VNF link re-mapping, and VNF state transfer. Then, we formalize the objective function in flow migration by considering the trade-off between load balancing and migration cost. Finally, we present the optimization problem formulation for flow migration.

3.2.1 Constraints

3.2.1.1 VNF Migration: VNF to NFV Node Re-mapping

For VNF to NFV node re-mapping in time interval k, VNF $V_h^{(r)}$ should be mapped to exactly one NFV node in $\mathcal{N}$, represented by

$$\sum_{n \in \mathcal{N}} \mathrm{x}_n^{rh}(k) = 1, \quad \forall (r, h) \in \mathcal{V}. \tag{3.2}$$

The physical locations of dummy VNFs, i.e., source and destination nodes, are fixed and confined by

$$\mathrm{x}_{\mathrm{n}_1^{(r)}}^{r0}(k) = 1, \qquad r \in \mathcal{R} \tag{3.3a}$$

$$\mathrm{x}_n^{r0}(k) = 0, \qquad r \in \mathcal{R}, n \in \mathcal{N} \cup \mathcal{A} \backslash \mathrm{n}_1^{(r)} \tag{3.3b}$$

$$\mathrm{x}_{\mathrm{n}_2^{(r)}}^{r(H_r+1)}(k) = 1, \quad r \in \mathcal{R} \tag{3.3c}$$

$$\mathrm{x}_n^{r(H_r+1)}(k) = 0, \quad r \in \mathcal{R}, n \in \mathcal{N} \cup \mathcal{A} \backslash \mathrm{n}_2^{(r)}. \tag{3.3d}$$

3.2.1.2 Resource Scaling

For vertical resource scaling during interval k, the processing rate in packet/s allocated to VNF $V_h^{(r)}$ by NFV node $n \in \mathcal{N}$ during interval k, i.e., $S_n^{rh}(k)$, is lower bounded by $\mathrm{x}_n^{rh}(k)\lambda^{(r)}(k)$ and upper bounded by $\frac{\mathrm{x}_n^{rh}(k)\mathsf{C}_n}{P_h^{(r)}b_r}$, represented as

$$\mathrm{x}_n^{rh}(k)\lambda^{(r)}(k) \leq S_n^{rh}(k) \leq \frac{\mathrm{x}_n^{rh}(k)\mathsf{C}_n}{P_h^{(r)}b_r}, \quad \forall (r,h) \in \mathcal{V}, \forall n \in \mathcal{N}. \tag{3.4}$$

Depending on whether VNF $V_h^{(r)}$ is placed at NFV node $n \in \mathcal{N}$ during interval k or not, the lower and upper bounds in (3.4) have different meanings. If $\mathrm{x}_n^{rh}(k) = 0$, constraint (3.4) is equivalent to $S_n^{rh}(k) = 0$, indicating that an NFV node allocates no computing resources to a VNF if the VNF is not mapped to it. If $\mathrm{x}_n^{rh}(k) = 1$, constraint (3.4) is equivalent to $\lambda^{(r)}(k) \leq S_n^{rh}(k) \leq \frac{\mathsf{C}_n}{P_h^{(r)}b_r}$. The left-hand-side guarantees the VNF processing queue stability for VNF $V_h^{(r)}$ at NFV node $n \in \mathcal{N}$. The right-hand-side guarantees that the allocated packet processing rate to VNF $V_h^{(r)}$ does not exceed the maximum packet processing rate supported by NFV node $n \in \mathcal{N}$. The processing density model is used to transform the CPU processing resource capacity in cycle/s to the maximum packet processing rate in packet/s.

3.2.1.3 Joint VNF Migration and Resource Scaling

The joint VNF migration and resource scaling decisions are both intra-SFC and inter-SFC coupled.

In terms of intra-SFC coupling, the VNF migration and resource scaling decisions for different VNFs in an SFC are coupled due to the E2E delay requirement. Let $\boldsymbol{d}(k) = \{d_n^{rh}(k)\}$ be a positive continuous decision variable set for interval k, with $d_n^{rh}(k)$ denoting the average (dummy) delay on the queue associated with VNF $V_h^{(r)}$ at NFV node n. With Poisson traffic arrival and exponential packet processing time, the processing system is an M/M/1 queue. Then, $d_n^{rh}(k)$ is determined by the joint VNF migration and resource scaling decisions of VNF $V_h^{(r)}$, given by

$$d_n^{rh}(k) = \frac{1}{S_n^{rh}(k) - \mathrm{x}_n^{rh}(k)\lambda^{(r)}(k) + \epsilon_1}, \quad \forall (r,h) \in \mathcal{V}, \forall n \in \mathcal{N} \tag{3.5}$$

where $d_n^{rh}(k)$ is a dummy delay only if $\mathrm{x}_n^{rh}(k) = 0$. There is an upper bound constraint for $d_n^{rh}(k)$, explicitly showing its relationship with $\mathrm{x}_n^{rh}(k)$:

$$0 < d_n^{rh}(k) \leq \mathrm{x}_n^{rh}(k)\mathsf{D}_r + \left[1 - \mathrm{x}_n^{rh}(k)\right]\frac{1}{\epsilon_1}, \quad \forall (r,h) \in \mathcal{V}, \forall n \in \mathcal{N}. \tag{3.6}$$

For QoS satisfaction, the E2E processing delay requirement of an SFC is satisfied as long as the aggregation of all VNF processing delays in an SFC does not exceed a specified upper bound. Specifically, for SFC $r \in \mathcal{R}$, we have

$$\sum_{h \in \mathcal{H}_r} \sum_{n \in \mathcal{N}} \mathrm{x}_n^{rh}(k) d_n^{rh}(k) \leq \mathsf{D}_r, \quad \forall r \in \mathcal{R}. \tag{3.7}$$

In terms of inter-SFC coupling, the VNF migration and resource scaling decisions for VNFs belonging to different SFCs are coupled, due to the CPU computing resource capacity constraints at the NFV nodes. Depending on the VNF migration decisions, each NFV node is re-mapped with a different group of VNFs belonging to different SFCs. For the group of VNFs placed at each NFV node, their resource scaling decisions are constrained by the resource capacity of the corresponding NFV node. For NFV node $n \in \mathcal{N}$, define a binary decision variable, $\mathrm{w}_n(k)$, with $\mathrm{w}_n(k) = 1$ if switching happens at NFV node n during interval k, and $\mathrm{w}_n(k) = 0$ otherwise. According to the CPU polling scheme in Sect. 3.1.3.3, switching happens at NFV node n if and only if at least two VNFs are mapped to NFV node n. The value of $\mathrm{w}_n(k)$ is confined by an inequality constraint, given by

$$\frac{\sum_{(r,h)\in\mathcal{V}} \mathrm{x}_n^{rh}(k) - 1}{|\mathcal{V}|} \leq \mathrm{w}_n(k) \leq \frac{\sum_{(r,h)\in\mathcal{V}} \mathrm{x}_n^{rh}(k)}{\mathsf{c}}, \quad \forall n \in \mathcal{N} \tag{3.8}$$

where c is an arbitrary value from $(1, 2]$. If there is no VNF placed at NFV node n during interval k, we have $\sum_{(r,h)\in\mathcal{V}} \mathrm{x}_n^{rh}(k) = 0$, and constraint (3.8) gives $-\frac{1}{|\mathcal{V}|} \leq \mathrm{w}_n(k) \leq 0$; if there is one VNF placed at NFV node n during interval k, we have $\sum_{(r,h)\in\mathcal{V}} \mathrm{x}_n^{rh}(k) = 1$, and constraint (3.8) gives $0 \leq \mathrm{w}_n(k) \leq \frac{1}{c} < 1$; if there is at least two VNF placed at NFV node n during interval k, we have $\sum_{(r,h)\in\mathcal{V}} \mathrm{x}_n^{rh}(k) \geq 2$, and constraint (3.8) gives $0 < \mathrm{w}_n(k) \leq \frac{2}{c}$. As $\mathrm{w}_n(k)$ is a binary variable, we conclude that the inequality constraint in (3.8) confines that $\mathrm{w}_n(k) = 1$ if and only if at least two VNFs are placed at NFV node n during interval k.

Let $\eta(k)$ denote the maximum NFV node utilization factor among all NFV nodes in $\mathcal{N}$. Assume that $\eta(k)$ should not exceed a predefined upper bound η_U ($0 < \eta_U \leq 1$). Then, for NFV node $n \in \mathcal{N}$, we have

$$\sum_{(r,h)\in\mathcal{V}} \left[\frac{P_h^{(r)} b_r S_n^{rh}(k)}{\mathsf{C}_n} + \frac{\mathrm{w}_n(k)\mathrm{x}_n^{rh}(k)\mathtt{W}}{T_n} \right] \leq \eta(k) \leq \eta_U, \quad \forall n \in \mathcal{N} \tag{3.9}$$

with the left-hand-side being the expression for the utilization factor of NFV node $n \in \mathcal{N}$ during interval k, i.e., $\eta_n(k)$. In the left-hand-side, $\sum_{(r,h)\in\mathcal{V}} \left[\frac{P_h^{(r)} b_r S_n^{rh}(k)}{\mathsf{C}_n} \right]$ denotes the ratio of useful CPU time allocated to VNFs in a CPU polling period of NFV node n during interval k, and $\sum_{(r,h)\in\mathcal{V}} \left[\frac{\mathrm{w}_n(k)\mathrm{x}_n^{rh}(k)\mathtt{W}}{T_n} \right]$ denotes the ratio of switching time overhead in a CPU polling period of NFV node n during interval k.

3.2.1.4 Subflow to Virtual Link Re-mapping

The subflow to virtual link re-mapping is a byproduct of VNF to NFV node re-mapping decisions, since the mapping between subflow and virtual link is uniquely determined by the locations of the corresponding upstream and downstream VNFs. Hence, the flow migration problem focuses on the service-level connectivity and performance. However, to maintain the infrastructure-level connectivity and performance, the required new virtual links should be provisioned by the infrastructure SDN controller, introducing signaling overhead. As defined in Sect. 3.1.4.2, $y_{nn'}^{rh}(k)$ is a binary decision variable denoting whether or not subflow $Y_h^{(r)}$ is mapped to an new virtual link between $n, n' \in \mathcal{N} \cup \mathcal{A}$ during interval k. With a given virtual resource pool topology, $\mathbf{y}(k) = \{y_{nn'}^{rh}(k)\}$ is uniquely determined by the VNF to NFV node mapping variables, $\mathbf{x}(k) = \{\mathrm{x}_n^{rh}(k)\}$.

There are two cases where no new virtual link is required for subflow $Y_h^{(r)}$:

- The upstream and downstream (dummy) VNFs of subflow $Y_h^{(r)}$ are mapped to two NFV nodes (or edge switches) between which there are existing direct virtual links;
- The upstream and downstream VNFs of subflow $Y_h^{(r)}$ are mapped to the same NFV node.

Specifically, if (dummy) VNF $V_h^{(r)}$ and (dummy) VNF $V_{h+1}^{(r)}$ are mapped to node $n \in \mathcal{N} \cup \mathcal{A}$ and node $n' \in \mathcal{N} \cup \mathcal{A} \backslash \{n\}$ between which no virtual link exists, $y_{nn'}^{rh}(k)$ is equal to 1, given by

$$y_{nn'}^{rh}(k) = \sum_{e \in \mathcal{E}} \left(1 - I_n^{e,1} I_{n'}^{e,2}\right) \mathrm{x}_n^{rh}(k) \mathrm{x}_{n'}^{r(h+1)}(k), \quad \forall r \in \mathcal{R}, \forall h \in \{0\} \cup \mathcal{H}_r,$$
$$\forall n \in \mathcal{N} \cup \mathcal{A}, \forall n' \in \mathcal{N} \cup \mathcal{A} \backslash \{n\}. \tag{3.10}$$

For virtual link $e \in \mathcal{E}$, we have $\left(1 - I_n^{e,1} I_{n'}^{e,2}\right) = 1$ if (1) node $n \in \mathcal{N} \cup \mathcal{A}$ is the staring point but node $n' \in \mathcal{N} \cup \mathcal{A} \backslash \{n\}$ is not the end point, i.e., $I_n^{e,1} = 1$ and $I_{n'}^{e,2} = 0$, or (2) node $n \in \mathcal{N} \cup \mathcal{A}$ is not the staring point but node $n' \in \mathcal{N} \cup \mathcal{A} \backslash \{n\}$ is the end point, i.e., $I_n^{e,1} = 0$ and $I_{n'}^{e,2} = 1$, or (3) node $n \in \mathcal{N} \cup \mathcal{A}$ is not the staring point and node $n' \in \mathcal{N} \cup \mathcal{A} \backslash \{n\}$ is not the end point, i.e., $I_n^{e,1} = 0$ and $I_{n'}^{e,2} = 0$. Hence, for virtual link $e \in \mathcal{E}$, $\left(1 - I_n^{e,1} I_{n'}^{e,2}\right) \mathrm{x}_n^{rh}(k) \mathrm{x}_{n'}^{r(h+1)}(k) = 1$ indicates that (dummy) VNF $V_h^{(r)}$ and (dummy) VNF $V_{h+1}^{(r)}$ are mapped to node $n \in \mathcal{N} \cup \mathcal{A}$ and node $n' \in \mathcal{N} \cup \mathcal{A} \backslash \{n\}$ but an existing virtual link $e \in \mathcal{E}$ is not between nodes n and n'. Also, if $n = n'$, we must have

$$y_{nn}^{rh}(k) = 0, \quad \forall r \in \mathcal{R}, \ \forall h \in \mathcal{H}_r \backslash \{H_r\}, \ \forall n \in \mathcal{N} \tag{3.11}$$

to represent that no new virtual link is required if both the upstream and downstream VNFs of subflow $Y_h^{(r)}$ are mapped to NFV node $n \in \mathcal{R}$.

3.2.1.5 VNF State Transfer

The state transfer is another byproduct of VNF to NFV node re-mapping decisions, since the associate VNF state should be transferred for each migrated VNF. From interval $(k-1)$ to interval k, the set representing VNF to NFV node mapping changes from $\mathbf{x}(k-1) = \{\mathrm{x}_n^{rh}(k-1)\}$ to $\mathbf{x}(k) = \{\mathrm{x}_n^{rh}(k)\}$, where $\mathbf{x}(k-1)$ is known in interval k. With the definition of $g_{n,n'}^{rh}(k)$ in Sect. 3.1.4.1, $g_{n,n'}^{rh}(k)$ also indicates whether state transfer happens between NFV node $n \in \mathcal{N}$ and NFV node $n' \in \mathcal{N}$ or not during interval k. With given values of $\mathbf{x}(k-1) = \{\mathrm{x}_n^{rh}(k-1)\}$, $\boldsymbol{g}(k) = \{g_{n,n'}^{rh}(k)\}$ is uniquely determined by the VNF to NFV node mapping variables, $\mathbf{x}(k) = \{\mathrm{x}_n^{rh}(k)\}$.

Only if the mapped NFV node for VNF $V_h^{(r)}$ changes from NFV node $n \in \mathcal{N}$ during interval $(k-1)$ to NFV node $n' \in \mathcal{N}\backslash\{n\}$ during interval k, i.e., $\mathrm{x}_n^{rh}(k-1) = \mathrm{x}_{n'}^{rh}(k) = 1$, a state transfer happens between NFV node $n \in \mathcal{N}$ and NFV node $n' \in \mathcal{N}\backslash\{n\}$ during interval k, given by

$$g_{n,n'}^{rh}(k) = \mathrm{x}_n^{rh}(k-1)\mathrm{x}_{n'}^{rh}(k), \quad \forall (r,h) \in \mathcal{V},\ \forall n \in \mathcal{N},\ \forall n' \in \mathcal{N}\backslash\{n\}. \tag{3.12}$$

If the mapped NFV node for VNF $V_h^{(r)}$ keep unchanged at NFV node $n \in \mathcal{N}$ from interval $(k-1)$ to interval k, i.e., $\mathrm{x}_n^{rh}(k-1) = \mathrm{x}_n^{rh}(k) = 1$, no state transfer happens, given by

$$g_{n,n}^{rh}(k) = 0, \quad \forall (r,h) \in \mathcal{V},\ \forall n \in \mathcal{N}. \tag{3.13}$$

To finish all state transfers within the maximal tolerable service downtime, sufficient transmission resources should be allocated to the state transfers. As defined in Sect. 3.1.4.3, $B_{n,n'}^{rh}(k)$ is the transmission resource overhead to transfer the state of VNF $V_h^{(r)}$ from NFV node $n \in \mathcal{N}$ to NFV node $n' \in \mathcal{N}$ during interval k. Only when a state transfer happens between NFV node $n \in \mathcal{N}$ to NFV node $n' \in \mathcal{N}$ during interval k, we have $B_{n,n'}^{rh}(k) > 0$; otherwise, we have $B_{n,n'}^{rh}(k) = 0$. Accordingly, we have

$$0 \le B_{n,n'}^{rh}(k) \le g_{n,n'}^{rh}(k)\mathbb{M}, \quad \forall (r,h) \in \mathcal{V},\ \forall n, n' \in \mathcal{N} \tag{3.14}$$

where $\mathbb{M}$ is a big-$\mathbb{M}$ constant.

Let $\boldsymbol{u}(k) = \{u_{n,n'}^{rh}(k)\}$ be a positive continuous decision variable set for interval k, with $u_{n,n'}^{rh}(k)$ denoting the (dummy) delay to transfer state of VNF $V_h^{(r)}$ from NFV node $n \in \mathcal{N}$ during interval $(k-1)$ to NFV node $n' \in \mathcal{N}$ during interval k. It follows that

$$u_{n,n'}^{rh}(k) = \frac{U_h^{(r)}(k)}{B_{n,n'}^{rh}(k) + \epsilon_1}, \quad \forall (r,h) \in \mathcal{V}, \ \forall n, n' \in \mathcal{N} \tag{3.15}$$

where $0 < \epsilon_1 \ll 1$ is a constant to avoid $u_{n,n'}^{rh}(k)$ being undetermined, and $u_{n,n'}^{rh}(k)$ is a dummy delay only if $g_{n,n'}^{rh}(k) = 0$. In the parallel state transfer scheme, the service downtime of SFC $r \in \mathcal{R}$ does not exceed the maximal tolerable service downtime Ω_r only if all state transfers in the SFC are finished within time limit Ω_r. Hence, $u_{n,n'}^{rh}(k)$ has an upper bound

$$0 < u_{n,n'}^{rh}(k) \leq g_{n,n'}^{rh}(k)\Omega_r + \left[1 - g_{n,n'}^{rh}(k)\right] \frac{U_h^{(r)}(k)}{\epsilon_1}, \quad \forall (r,h) \in \mathcal{V}, \ \forall n, n' \in \mathcal{N}. \tag{3.16}$$

If $g_{n,n'}^{rh}(k) = 1$, the upper bound is the corresponding maximal tolerable service downtime Ω_r; otherwise, it is $\frac{U_h^{(r)}(k)}{\epsilon_1}$.

3.2.2 *Optimization Problem*

To support flow migration, the migration cost is composed of two parts: the total transmission resource overhead incurred by state transfers, and the total signaling overhead for configuring new virtual links required for flow rerouting. The latter is linear with the total number of new virtual links required for flow rerouting.

If all VNF states have the same size and all services have the same downtime limits, the amount of transmission resources required by the parallel state transfers linearly varies with the total number of state transfers, i.e., the number of VNF migrations. Similarly, if more VNFs are migrated to alternative NFV nodes, more subflows are re-mapped to alternative virtual links, which increases the chance of requiring new virtual links. Hence, the migration cost minimization leads to limited modification to the original VNF to NFV node mapping. However, in this case, the loads on different NFV nodes can be imbalanced, which can result in more migrations in the subsequent time intervals. A balanced load distribution makes the network more tolerant of future demand changes, which is beneficial for achieving long-term efficient resource utilization [15].

Therefore, we formulate the flow migration problem as an optimization problem, by jointly considering the migration cost and load balancing. The goal is to achieve load balancing among NFV nodes, with minimal migration cost. Let $O(k)$ denote the objective function for time interval k, which is given by

$$O(k) = \omega_1 \eta(k) + \omega_2 \sum_{(r,h)\in\mathcal{V}} \sum_{n,n'\in\mathcal{N}} \frac{B^{rh}_{n,n'}(k)}{B_{min}} + \omega_3 \sum_{r\in\mathcal{R}} \sum_{h\in\{0\}\cup\mathcal{H}_r} \sum_{n,n'\in\mathcal{N}\cup\mathcal{A}} y^{rh}_{nn'}(k) \quad (3.17)$$

where $\omega_1, \omega_2, \omega_3$ are tunable weights to control the priority of the three components, with $\omega_1 + \omega_2 + \omega_3 = 1$. In the right hand side of (3.17), the first term is the cost for imbalanced loading among NFV nodes since minimizing $\eta(k)$ achieves load balancing among all the NFV nodes, the second term is the cost for the overall normalized transmission resource overhead due to state transfers with a normalization ratio of $\frac{1}{B_{min}}$, the third term is the cost for new virtual links required by flow rerouting. The normalization makes the three components in objective function (3.17) comparable, based on which $\omega_1, \omega_2, \omega_3$ can be selected on the same order of magnitude. A component in (3.17) is ignored if the corresponding weight is set to 0. If all weights are positive, all components in (3.17) are jointly optimized.

Then, the optimization problem is formulated as

$$\min_{\substack{\eta(k),\boldsymbol{B}(k),\boldsymbol{y}(k),\mathbf{x}(k),\boldsymbol{g}(k),\\ \boldsymbol{u}(k),\boldsymbol{S}(k),\mathbf{w}(k),\boldsymbol{d}(k)}} \quad O(k) \quad (3.18a)$$

$$\text{s.t.} \quad (3.2) - (3.16) \quad (3.18b)$$

$$\boldsymbol{d}(k), \boldsymbol{u}(k) > 0 \quad (3.18c)$$

$$\mathbf{x}(k), \mathbf{w}(k), \boldsymbol{g}(k), \boldsymbol{y}(k) \in \{0, 1\}. \quad (3.18d)$$

Remark 3.2 The optimization problem in (3.18) is non-convex due to quadratic constraints (3.5), (3.7), (3.9), (3.10), and (3.15). In Sect. 3.3, a mixed integer quadratically constraint programming (MIQCP) problem is formulated through constraint transformations, and the relationship between the optimal solutions to the original problem in (3.18) and the MIQCP problem is analyzed. An optimum mapping algorithm is presented, which describes how to obtain an optimum of the original problem in (3.18) from an MIQCP optimum.

The diagram in Fig. 3.8 summarizes the main decision variables, constraints, and costs in the flow migration optimization model. In the figure, the decision variables, constraints, and costs related to joint VNF migration and resource scaling, subflow to virtual link re-mapping, and VNF state transfer are indicated by red, green, and blue colors, respectively.

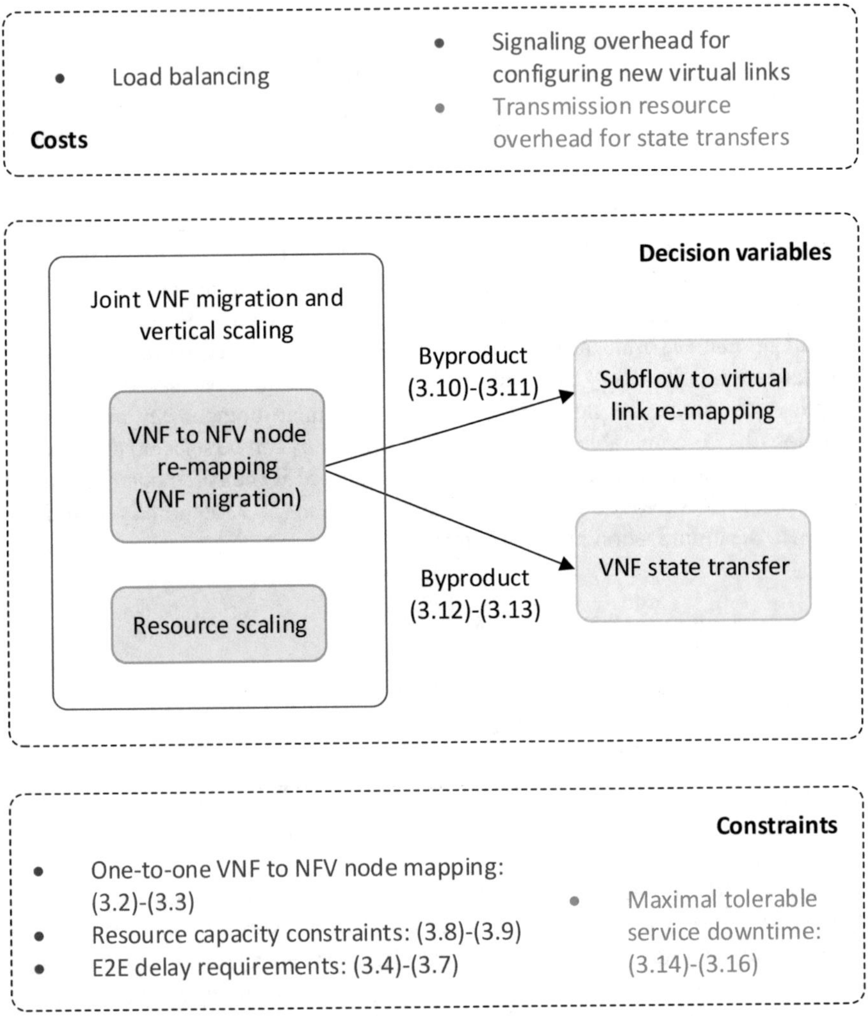

Fig. 3.8 Diagram of the flow migration optimization model

3.3 Optimal MIQCP Solution

In this section, an MIQCP optimal solution to the optimization problem is presented, which is based on MIQCP problem transformation and optimum mapping. A diagram of the optimal MIQCP solution is shown in Fig. 3.9.

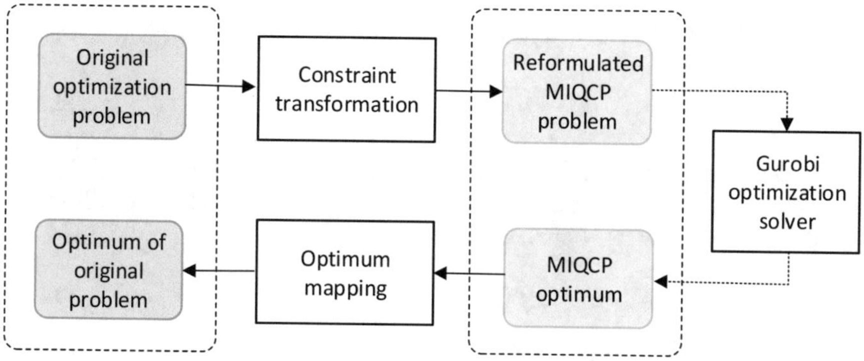

Fig. 3.9 Diagram of the optimal MIQCP solution

3.3.1 Constraint Transformation

The quadratic constraints (3.7), (3.9), and (3.10) in the original optimization problem in (3.18) can be transformed to equivalent linear forms using the big-$\mathbb{M}$ method. Quadratic constraints (3.5) and (3.15) cannot be linearized due to product terms of two continuous variables, but they can be replaced by combinations of linear constraints and rotated quadratic cone constraints. The new problem with transformed and replaced constraints is an MIQCP problem, which is not equivalent to the original problem. In this section, we discuss the relationship between the two problems.

By introducing an auxiliary nonnegative continuous decision variable set, $\boldsymbol{\iota}(k) = \{\iota_n^{rh}(k)\}$, we linearize constraint (3.7) based on the big-$\mathbb{M}$ method with $\mathbb{M} = \frac{1}{\epsilon_1}$ as

$$\sum_{h \in \mathcal{H}_r} \sum_{n \in \mathcal{N}} \iota_n^{rh}(k) \le \mathsf{D}_r, \quad \forall r \in \mathcal{R} \tag{3.19a}$$

$$d_n^{rh}(k) - \frac{1}{\epsilon_1}\left[1 - \mathrm{x}_n^{rh}(k)\right] \le \iota_n^{rh}(k) \le d_n^{rh}(k), \quad \forall (r,h) \in \mathcal{V}, \forall n \in \mathcal{N} \tag{3.19b}$$

$$0 \le \iota_n^{rh}(k) \le \frac{1}{\epsilon_1}\mathrm{x}_n^{rh}(k), \quad \forall (r,h) \in \mathcal{V}, \forall n \in \mathcal{N}. \tag{3.19c}$$

By introducing an auxiliary nonnegative continuous decision variable set, $\boldsymbol{\zeta}(k) = \{\zeta_n(k)\}$, we linearize constraint (3.9) based on the big-$\mathbb{M}$ method with $\mathbb{M} = |\mathcal{V}|$, given by

$$\sum_{(r,h) \in \mathcal{V}} \frac{P_h^{(r)} b_r S_n^{rh}(k)}{\mathsf{C}_n} T_n + \zeta_n(k)\mathtt{W} \le \eta(k) T_n \le \eta_U T_n, \quad \forall n \in \mathcal{N} \tag{3.20a}$$

$$\sum_{(r,h)\in\mathcal{V}} \mathrm{x}_n^{rh}(k) - |\mathcal{V}|\left[1 - \mathrm{w}_n(k)\right] \leq \zeta_n(k) \leq \sum_{(r,h)\in\mathcal{V}} \mathrm{x}_n^{rh}(k), \quad \forall n \in \mathcal{N} \tag{3.20b}$$

$$0 \leq \zeta_n(k) \leq |\mathcal{V}| \mathrm{w}_n(k), \quad \forall n \in \mathcal{N}. \tag{3.20c}$$

By introducing an auxiliary binary decision variable set, $\boldsymbol{\xi}(k) = \{\xi_{nn'}^{rh}(k)\}$, we get an equivalent linear form of constraint (3.10) for $\forall r \in \mathcal{R}, \forall h \in \{0\} \cup \mathcal{H}_r$, $\forall n \in \mathcal{N} \cup \mathcal{A}$, and $\forall n' \in \mathcal{N} \cup \mathcal{A}$, given by

$$\xi_{nn'}^{rh}(k) \leq \mathrm{x}_n^{rh}(k) \tag{3.21a}$$

$$\xi_{nn'}^{rh}(k) \leq \mathrm{x}_{n'}^{r(h+1)}(k) \tag{3.21b}$$

$$\xi_{nn'}^{rh}(k) \geq \mathrm{x}_n^{rh}(k) + \mathrm{x}_{n'}^{r(h+1)}(k) - 1 \tag{3.21c}$$

$$y_{nn'}^{rh}(k) = \sum_{e\in\mathcal{E}} \left(1 - I_n^{e,1} I_{n'}^{e,2}\right) \xi_{nn'}^{rh}(k). \tag{3.21d}$$

With linearized constraints (3.19),(3.20) and (3.21), the problem in (3.18) can be transformed to an MIQCP problem, if constraint (3.5) is replaced by

$$\varpi_n^{rh}(k) = S_n^{rh}(k) - \mathrm{x}_n^{rh}(k)\lambda^{(r)}(k) + \epsilon_1, \quad \forall (r,h) \in \mathcal{V}, \forall n \in \mathcal{N} \tag{3.22a}$$

$$\varpi_n^{rh}(k) \geq \epsilon_1, \qquad \forall (r,h) \in \mathcal{V}, \forall n \in \mathcal{N} \tag{3.22b}$$

$$d_n^{rh}(k)\varpi_n^{rh}(k) \geq \Xi^2, \quad \forall (r,h) \in \mathcal{V}, \forall n \in \mathcal{N} \tag{3.22c}$$

$$\Xi = 1 \tag{3.22d}$$

where $\boldsymbol{\varpi}(k) = \{\varpi_n^{rh}(k)\}$ is an auxiliary continuous decision variable set and Ξ is an auxiliary continuous decision variable, and if constraint (3.15) is replaced by

$$\varrho_{n,n'}^{rh}(k) = B_{n,n'}^{rh}(k) + \epsilon_1, \qquad \forall (r,h) \in \mathcal{V},\ n,n' \in \mathcal{N} \tag{3.23a}$$

$$\varrho_{n,n'}^{rh}(k) \geq \epsilon_1, \qquad \forall (r,h) \in \mathcal{V},\ n,n' \in \mathcal{N} \tag{3.23b}$$

$$u_{n,n'}^{rh}(k)\varrho_{n,n'}^{rh}(k) \geq \varsigma_h^{(r)}(k)^2, \quad \forall (r,h) \in \mathcal{V},\ n,n' \in \mathcal{N} \tag{3.23c}$$

$$\varsigma_h^{(r)}(k) = \sqrt{U_h^{(r)}(k)}, \qquad \forall (r,h) \in \mathcal{V} \tag{3.23d}$$

where $\boldsymbol{\varrho}(k) = \{\varrho_{n,n'}^{rh}(k)\}$ and $\boldsymbol{\varsigma}(k) = \{\varsigma_h^{(r)}(k)\}$ are auxiliary continuous decision variable sets. The MIQCP problem is formulated as

$$\min_{\substack{\eta(k),\boldsymbol{B}(k),\boldsymbol{y}(k),\mathbf{x}(k),\boldsymbol{g}(k),\boldsymbol{u}(k)\\ \boldsymbol{S}(k),\mathbf{w}(k),\boldsymbol{d}(k),\iota(k),\boldsymbol{\zeta}(k),\boldsymbol{\xi}(k)}} O(k) \tag{3.24a}$$

$$\text{s.t.} \quad (3.2)\text{–}(3.4), (3.6), (3.8), (3.11)\text{–}(3.14) \tag{3.24b}$$

$$(3.16), (3.19)\text{–}(3.22) \tag{3.24c}$$

$$\boldsymbol{d}(k), \boldsymbol{u}(k) > 0 \tag{3.24d}$$

$$\mathbf{x}(k), \mathbf{w}(k), \boldsymbol{g}(k), \boldsymbol{y}(k), \boldsymbol{\xi}(k) \in \{0, 1\}. \tag{3.24e}$$

The MIQCP problem is solvable in Gurobi optimization solver. Its objective function and all its constraints except constraints (3.22c) and (3.23c) are linear. Constraints (3.22c) and (3.23c) are rotated second-order (quadratic) cone constraints which are supported in Gurobi solver. In Sect. 3.3.2, we discuss the relationship between the optimal solutions to the original problem in (3.18) and the reformulated MIQCP problem in (3.24), and prove the zero optimality gap between the two problems.

3.3.2 Optimality Gap and Optimum Mapping

Let the Gurobi optimization solver solve the reformulated MIQCP problem in (3.24), which gives the optimal values of all the decision variables including the auxiliary decision variables. The group of optimal decision variables excluding the optimal auxiliary decision variables, i.e., $\eta^\star(k)$, $\boldsymbol{B}^\star(k)$, $\boldsymbol{y}^\star(k)$, $\mathbf{x}^\star(k)$, $\boldsymbol{g}^\star(k)$, $\boldsymbol{u}^\star(k)$, $\boldsymbol{S}^\star(k)$, $\mathbf{w}^\star(k)$, $\boldsymbol{d}^\star(k)$, is referred to as an MIQCP optimum (indicated by "$\star$"). With an MIQCP optimum, the optimum of the original problem in (3.18) (indicated by "$*$"), i.e., $\eta^*(k)$, $\boldsymbol{B}^*(k)$, $\boldsymbol{y}^*(k)$, $\mathbf{x}^*(k)$, $\boldsymbol{g}^*(k)$, $\boldsymbol{u}^*(k)$, $\boldsymbol{S}^*(k)$, $\mathbf{w}^*(k)$, $\boldsymbol{d}^*(k)$, can be obtained with Proposition 3.1.

Proposition 3.1 *The optimality gap between the original problem in* (3.18) *and the reformulated MIQCP problem in* (3.24) *is zero, i.e., an optimum of the original problem in* (3.18) *is either a unique optimum or one of multiple optimal solutions to the MIQCP problem. Given an MIQCP optimum, an optimum of the original problem in* (3.18) *can be obtained by an optimum mapping algorithm in Algorithm 3.1.*

Proof The fundamental difference between the MIQCP problem and the original problem in (3.18) lies in "$\geq$" signs in rotated quadratic cone constraints (3.22c) and (3.23c). If both constraints are active in an MIQCP optimum, i.e., all the "$\geq$" signs achieve equality, the MIQCP optimum is also an optimum of the original problem in (3.18). If either rotated quadratic cone constraint is inactive in an MIQCP optimum, i.e., the "$\geq$" sign achieves "$>$", the MIQCP optimum is not the optimum of the original problem. Next, let's prove that the optimum of the original problem in (3.18) can be calculated from the MIQCP optimum, if either rotated quadratic cone constraint is inactive.

First, assume that there is an inactive constraint (3.22c) in an MIQCP optimum, then $d_n^{rh\star}(k)\left[S_n^{rh\star}(k) - \mathrm{x}_n^{rh\star}(k)\lambda^{(r)}(k) + \epsilon_1\right] > 1$. If $\mathrm{x}_n^{rh\star}(k) = 0$, we must have $S_n^{rh\star}(k) = 0$ according to (3.4) and $d_n^{rh\star}(k)$ being a dummy delay. Then, it does not affect the objective value. Thus, we consider only the constraint with $\mathrm{x}_n^{rh\star}(k) = 1$. For NFV node $n \in \mathcal{N}$, it is referred to as the dominating NFV node if the utilization factor is equal to the maximum NFV node utilization factor, i.e., $\eta_n{}^\star(k) = \eta^\star(k)$. If $\eta_n{}^\star(k) < \eta^\star(k)$, NFV node $n \in \mathcal{N}$ is referred to as a non-dominating NFV node. Depending on ω_1 and $\eta_n{}^\star(k)$, there are four cases:

- **Case 1**: $\omega_1 > 0$, and NFV node n is the only dominating NFV node, i.e., $\eta_n{}^\star(k) = \eta^\star(k)$ and $\eta_{n'}{}^\star(k) < \eta^\star(k), \forall n' \in \mathcal{N}\backslash\{n\}$. Then, if $S_n^{rh\star}(k)$ is replaced by $S_n^{rh\circ}(k)$, with $S_n^{rh\circ}(k) < S_n^{rh\star}(k)$ and $d_n^{rh\star}(k)[S_n^{rh\circ}(k) - \mathrm{x}_n^{rh\star}(k)\lambda^{(r)}(k) + \epsilon_1] = 1$, all constraints are satisfied but $\eta^\star(k)$ can be further reduced. Hence, the assumption must be false. In other words, for the single dominating NFV node, constraint (3.22c) must be active, and $[d_n^{rh\star}(k), S_n^{rh\star}(k)]$ is an optimal pair in the optimum of the original problem, i.e., $[d_n^{rh*}(k), S_n^{rh*}(k)] = [d_n^{rh\star}(k), S_n^{rh\star}(k)]$;
- **Case 2**: $\omega_1 > 0$ and NFV node n is a non-dominating NFV node, i.e., $\eta_n{}^\star(k) < \eta^\star(k)$. Then, if $d_n^{rh\star}(k)$ is replaced by $d_n^{rh\circ}(k)$, with $d_n^{rh\circ}(k) < d_n^{rh\star}(k)$ and $d_n^{rh\circ}(k)[S_n^{rh\star}(k) - \mathrm{x}_n^{rh\star}(k)\lambda^{(r)}(k) + \epsilon_1] = 1$, all constraints are satisfied with the objective value unchanged, as $\eta^\star(k)$ is determined by the dominating NFV node. Thus, $[d_n^{rh\circ}(k), S_n^{rh\star}(k)]$ is an optimal pair in another MIQCP optimum, which is also the optimal pair in the optimum of the original problem, i.e., $[d_n^{rh*}(k), S_n^{rh*}(k)] = [d_n^{rh\circ}(k), S_n^{rh\star}(k)]$;
- **Case 3**: $\omega_1 > 0$, and there are multiple dominating NFV nodes including NFV node n with utilization factor $\eta^\star(k)$. There must be at least one dominating NFV node satisfying an active constraint (3.22c). Otherwise, the objective value can be further reduced according to Case 1. Hence, for each dominating NFV node with active constraint (3.22c), $[d_n^{rh\star}(k), S_n^{rh\star}(k)]$ is an optimal pair in the optimum of the original problem, i.e, $[d_n^{rh*}(k), S_n^{rh*}(k)] = [d_n^{rh\star}(k), S_n^{rh\star}(k)]$. For each dominating NFV node with inactive constraint (3.22c), $[d_n^{rh\circ}(k), S_n^{rh\star}(k)]$ is an optimal pair in another MIQCP optimum, which is also the optimal pair in the optimum of the original problem, i.e., $[d_n^{rh*}(k), S_n^{rh*}(k)] = [d_n^{rh\circ}(k), S_n^{rh\star}(k)]$;
- **Case 4**: $\omega_1 = 0$, and $\eta(k)$ is not optimized. If we replace $S_n^{rh\star}(k)$ by $S_n^{rh\circ}(k)$, all constraints are satisfied and the objective value is unchanged. Thus, $[d_n^{rh\star}(k), S_n^{rh\circ}(k)]$ is an optimal pair in another MIQCP optimum, which is also the optimal pair in the optimum of the original problem, i.e., $[d_n^{rh*}(k), S_n^{rh*}(k)] = [d_n^{rh\star}(k), S_n^{rh\circ}(k)]$.

Second, assume that there is an inactive constraint (3.23c) in an MIQCP optimum, i.e., $u_{n,n'}^{rh\ \star}(k)\left[B_{n,n'}^{rh\ \star}(k) + \epsilon_1\right] > U_h^{(r)}(k)$. Similarly, if $B_{n,n'}^{rh\ \star}(k) = 0$, it does not affect the objective value. Thus, we consider only the case with $B_{n,n'}^{rh\ \star}(k) > 0$. If $B_{n,n'}^{rh\ \star}(k)$ is replaced by $B_{n,n'}^{rh\ \circ}(k)$, with $B_{n,n'}^{rh\ \circ}(k) < B_{n,n'}^{rh\ \star}(k)$ and $u_{n,n'}^{rh\ \star}(k)\left[B_{n,n'}^{rh\ \circ}(k) + \epsilon_1\right] = U_h^{(r)}(k)$, all constraints are still satisfied. The objective

Algorithm 3.1: Post-processing to MIQCP optimum

1 **Input:** $\eta^\star(k)$, $\boldsymbol{B}^\star(k)$, $\boldsymbol{y}^\star(k)$, $\mathbf{x}^\star(k)$, $\boldsymbol{g}^\star(k)$, $\boldsymbol{u}^\star(k)$, $\boldsymbol{S}^\star(k)$, $\mathbf{w}^\star(k)$, $\boldsymbol{d}^\star(k)$.
2 *Initialization* $(* = \star)$.
3 **for** $(r, h) \in \mathcal{V}$, $n \in \mathcal{N}$ **do**
4   **if** $d_n^{rh\star}(k)\,\varpi_n^{rh\star}(k) > (c^\star)^2$ *and* $\mathrm{x}_n^{rh\star}(k) == 1$, **then**
5   **if** $\omega_1 > 0$, **then**
6             $d_n^{rh*}(k) = \frac{1}{S_n^{rh\star}(k) - \mathrm{x}_n^{rh\star}(k)\,\lambda^{(r)}(k) + \epsilon_1}$
7   **if** $\omega_1 == 0$, **then**
8             $S_n^{rh*}(k) = \frac{1}{d_n^{rh\star}(k)} + \mathrm{x}_n^{rh\star}(k)\,\lambda^{(r)}(k) - \epsilon_1$

9 **for** $(r, h) \in \mathcal{V}$, $n, n' \in \mathcal{N}$ **do**
10   **if** $u_{n,n'}^{rh\;\star}(k)\,\varrho_{n,n'}^{rh\;\star}(k) > (\varsigma_h^{(r)}(k)^\star)^2$ *and* $B_{n,n'}^{rh\;\star}(k) > 0$, **then**
        $B_{n,n'}^{rh\;*}(k) = \frac{U_h^{(r)}(k)}{u_{n,n'}^{rh\;\star}(k)} - \epsilon_1$
11 **if** $\omega_1 == 0$, **then** *calculate* $\eta^*(k)$
12 **Output:** $\eta^*(k)$, $\boldsymbol{B}^*(k)$, $\boldsymbol{y}^*(k)$, $\mathbf{x}^*(k)$, $\boldsymbol{g}^*(k)$, $\boldsymbol{u}^*(k)$, $\boldsymbol{S}^*(k)$, $\mathbf{w}^*(k)$, $\boldsymbol{d}^*(k)$.

value is unchanged if $\omega_2 = 0$, in which case $[u_{n,n'}^{rh\;\star}(k), B_{n,n'}^{rh\;\circ}(k)]$ is an optimal pair in another MIQCP optimum. Otherwise ($\omega_2 > 0$), the objective value can be further reduced, inferring that the assumption must be false.

In summary, an MIQCP optimum with inactive constraints in (3.22c) and (3.23c) can be mapped to another MIQCP optimum with active constraints in (3.22c) and (3.23c), without affecting other constraints and the objective value. The mapped MIQCP optimum is also the optimum of the original problem in (3.18). The mapping algorithm between the optimal solutions is provided in Algorithm 3.1. In the algorithm, Line 2 initializes the optimum of the original problem as the MIQCP optimum. Lines 3–8 deal with inactive constraint (3.22c). Lines 5–6 correspond to the non-dominating NFV nodes in Case 2 and the dominating NFV nodes in Case 3, both with inactive constraint (3.22c). Lines 9–10 deal with inactive constraint (3.23c), corresponding to $\omega_2 = 0$.

3.4 Low-Complexity Heuristic Flow Migration Algorithm

Although the original problem in (3.18) can be solved by the optimal MIQCP solution according to Proposition 3.1, the computational time is high due to NP-hardness of the MIQCP problem.

Remark 3.3 The MIQCP problem is NP-hard.

Proof To prove the NP-hardness, it is sufficient to consider a special case in which services with $\mathsf{D}_r \to \infty$ are embedded in a fully-connected virtual resource pool. We also consider zero VNF state size, zero switching time overhead, and sufficiently large computing resource capacity for each NFV node holding all VNFs without overloading [16, 17]. In such a case, the MIQCP problem can be reduced from a multiprocessor scheduling problem [18]. The multiprocessor scheduling problem minimizes the maximum load among a number of processors which are assigned with a number of tasks with different loads, which is proved to be NP-hard.

For time tractability, a low-complexity modular heuristic algorithm is used to obtain a suboptimal solution to the original problem in (3.18). We consider only the case where all components in objective function (3.17) are jointly optimized, i.e., $\omega_1, \omega_2, \omega_3 > 0$. In this case, we assume that one VNF migration is penalized more than imbalanced loading (i.e., $\eta(k)$ reaching its upper bound η_U). Then, the condition of $\omega_1 \eta_U < \omega_2$ should be satisfied in the worst case, if all VNF migrations incur the same transmission resource overhead for state transfer and require no new virtual links for flow rerouting.

Accordingly, in the heuristic algorithm, we first minimize the number of overloaded NFV nodes with utilization factors greater than η_U, and make migration decisions at overloaded NFV nodes, after which $\eta(k)$ is equal to η_U. Afterwards, $\eta(k)$ is further reduced for load balancing. The algorithm is insensitive to ω_1 but sensitive to $\frac{\omega_2}{\omega_3}$, due to migration cost aware migration decisions. Therefore, it provides a sub-optimal solution to the problem in (3.18) with $\omega_1 \eta_U < \omega_2$.

3.4.1 Algorithm Overview

A flowchart of the heuristic algorithm is given in Fig. 3.10. The heuristic algorithm is to determine a joint VNF migration and resource scaling plan for interval k in the presence of predicted traffic variations, i.e., from $\{\lambda^{(r)}(k-1)\}$ to $\{\lambda^{(r)}(k)\}$.

At the end of interval $k-1$, the VNF to NFV node mapping results are known, denoted by $\mathbf{x}(k-1) = \{\mathrm{x}_n^{rh}(k-1)\}$. Let $\mathrm{x}_n^{rh}(k) = \mathrm{x}_n^{rh}(k-1)$ be the initial mapping between VNF $V_h^{(r)}$ to NFV node $n \in \mathcal{N}$ for interval k. Consider that the E2E (processing) delay requirement for each SFC is initially decomposed into a set of per-hop processing delay requirements at the VNFs (i.e., hop delay bounds), denoted by $\{\mathsf{D}^{rh}(k), \forall (r,h) \in \mathcal{V}\}$, with $\sum_{h \in \mathcal{H}_r} \mathsf{D}^{rh}(k) = \mathsf{D}_r$ for SFC $r \in \mathcal{R}$. Let $\mathsf{D}^{rh}(k) = \sum_{n \in \mathcal{N}} \mathrm{x}_n^{rh}(k-1) d_n^{rh}(k-1)$ be the initial hop delay bound at VNF $V_h^{(r)}$ for interval k.

With given VNF to NFV node mapping and hop delay bounds for interval k, we first calculate NFV node utilization factor $\eta_n(k)$ for each NFV node $n \in \mathcal{N}$ during interval k, with the predicted traffic rates $\{\lambda^{(r)}(k)\}$, based on the M/M/1 queue delay model and processing density model. Specifically, for NFV node $n \in \mathcal{N}$, we have

$$\eta_n(k) = \sum_{(r,h) \in \mathcal{V}} \left(\frac{P_h^{(r)} b_r S_n^{rh}(k)}{\mathsf{C}_n} + \frac{\mathrm{w}_n(k) \mathrm{x}_n^{rh}(k) \mathtt{W}}{T_n} \right) \tag{3.25}$$

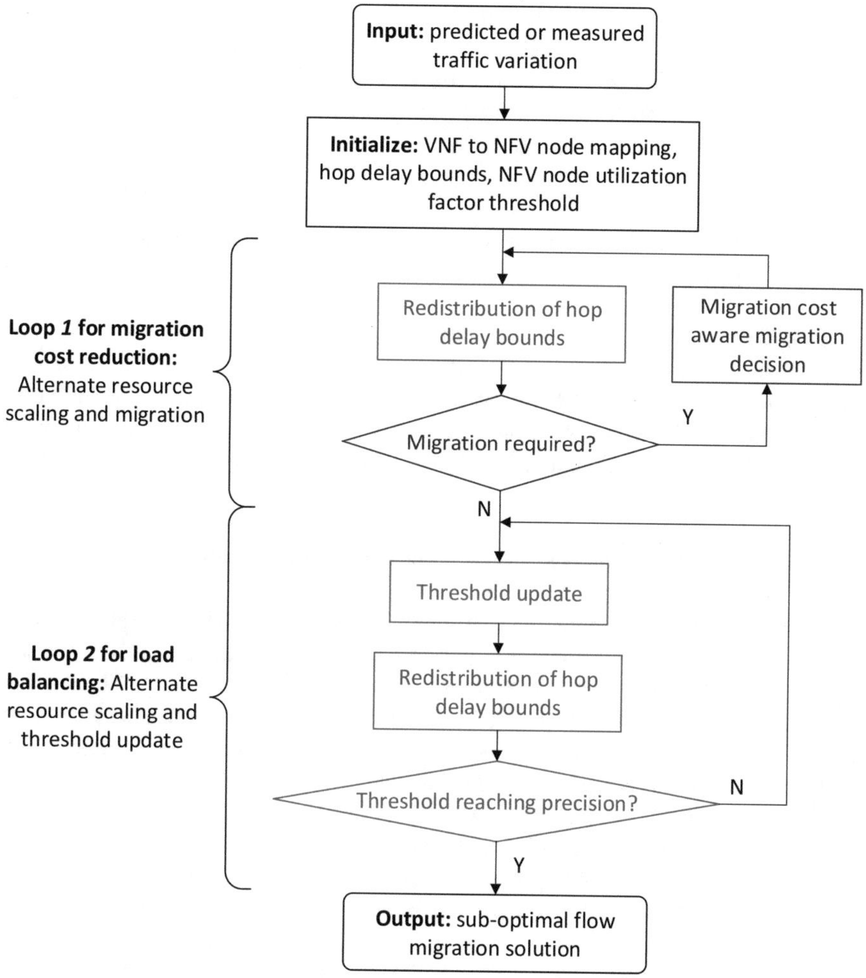

Fig. 3.10 Flowchart of the heuristic algorithm for dynamic flow migration

where $\mathrm{w}_n(k)$ is calculated from (3.8) and $S_n^{rh}(k)$ is given by

$$S_n^{rh}(k) = \left(\lambda^{(r)}(k) + \frac{1}{\mathsf{D}^{rh}(k)}\right) \mathrm{x}_n^{rh}(k). \tag{3.26}$$

Due to traffic variations, the calculated NFV node utilization factors can be larger than the predefined upper bound η_U of the maximum NFV node utilization factor, and can be even larger than 1, which indicates resource overloading or resource capacity constraint violation. Define threshold η_{th} for the NFV node utilization factors, beyond which an NFV node is considered as overloaded. Let η_U be the initial value of η_{th}.

The overloading status of NFV nodes depends on three factors: VNF to NFV node mapping, hop delay bounds, and NFV node utilization factor threshold. By comparing the calculated NFV node utilization factors with threshold η_{th}, a set of overloaded NFV nodes is identified as potential bottlenecks, where potential VNF migrations and resource scaling are required.

3.4.1.1 Migration Cost Reduction: Alternate Resource Scaling and Migration

Even if potential bottlenecks are identified, migration may not be necessary. For a given threshold, η_{th}, how an E2E delay requirement is decomposed into hop delay bounds affects the number of overloaded NFV nodes. To reduce the migration cost, we first perform resource scaling for each potential bottleneck NFV node, by adjusting the hop delay bounds of the VNFs at the NFV nodes.

Resource Scaling Through Redistribution of Hop Delay Bounds

By making hop delay bounds less stringent on overloaded NFV nodes and more stringent on underloaded ones, it is possible to reduce the number of overloaded NFV nodes. The basic idea is as follows: if an SFC traverses both overloaded and underloaded NFV nodes, utilization factors of the underloaded ones are increased to η_{th}, by reducing corresponding hop delay bounds, and utilization factors of the overloaded ones are decreased, by increasing corresponding hop delay bounds. This strategy is referred to as delay scaling, which is performed iteratively until there is no SFC traversing both overloaded and underloaded NFV nodes. The iterative delay scaling procedure with given threshold, η_{th}, is referred to as the redistribution of hop delay bounds. More details on redistribution of hop delay bounds are given in Sect. 3.4.2.

As illustrated in Fig. 3.10, a redistribution of hop delay bounds with the initial threshold ($\eta_{th} = \eta_U$) is performed after the initialization step. If the number of overloaded NFV nodes is reduced to zero, no migration is required. Otherwise, migration is necessary to overcome resource overloading.

Migration Decisions

Migration decisions are made sequentially, i.e., only a pair of variables in set $\{\mathrm{x}_n^{rh}(k)\}$ is updated in one migration decision. Each migration decision is followed by a redistribution of hop delay bounds, until no more migration is required, as illustrated by Loop 1 in Fig. 3.10.

A migration cost aware VNF migration decision includes three steps, i.e., identification of a bottleneck NFV node, selection of an SFC to migrate, and selection of a target NFV node. First, the most heavily loaded NFV node is identified as the bottleneck NFV node. Next, an SFC to migrate from the bottleneck NFV node and a target NFV node to accommodate the migrated SFC are jointly selected to minimize the migration cost. If there are multiple choices, an SFC with the largest resource demand is migrated to the closest target NFV node. More details on migration decision are discussed in Sect. 3.4.3.

With alternate resource scaling and migration, migration cost is greedily reduced with the potential reduction of overloaded NFV nodes by redistribution of hop delay bounds, and the migration cost awareness in migration decision.

3.4.1.2 Load Balancing: Alternate Resource Scaling and Threshold Update

If no potential bottlenecks are detected after the initial redistribution of hop delay bounds, or if all the detected potential bottlenecks are removed by alternate resource scaling and migration in Loop 1, load balancing is the only remaining objective. Then, as illustrated in the flowchart, in the second loop, the NFV node utilization factors are gradually balanced by iterative redistribution of hop delay bounds with threshold updating. The threshold, η_{th}, is updated from binary search, until it reaches sufficient precision. More details on threshold updating are discussed in Sect. 3.4.4.

3.4.2 *Redistribution of Hop Delay Bounds*

With the calculated NFV node utilization factors, $\{\eta_n(k), \forall n \in \mathcal{N}\}$, and the given NFV node utilization factor threshold, η_{th}, three sets of NFV nodes are identified based on their overloading status: $\mathcal{N}_O = \{n \in \mathcal{N} \,|\, \eta_n(k) > \eta_{th}\}$ consisting of overloaded NFV nodes, $\mathcal{N}_U = \{n \in \mathcal{N} \,|\, \eta_n(k) < \eta_{th}\}$ for underloaded NFV nodes, and $\mathcal{N}_E = \{n \in \mathcal{N} \,|\, \eta_n(k) = \eta_{th}\}$. Let binary variable $X_n^{(r)}(k)$ indicate whether SFC $r \in \mathcal{R}$ traverses NFV node $n \in \mathcal{N}$ during interval k, with $X_n^{(r)}(k) = 1$ if $\sum_{h \in \mathcal{H}_r} \mathrm{x}_n^{rh}(k) > 0$, and $X_n^{(r)}(k) = 0$ otherwise.

Let $\mathtt{f}_1^{(r)}$ be a binary flag indicating whether SFC $r \in \mathcal{R}$ traverses any overloaded NFV nodes in $\mathcal{N}_O$, with $\mathtt{f}_1^{(r)} = 1$ if $\sum_{n \in \mathcal{N}_O} X_n^{(r)}(k) > 0$, and $\mathtt{f}_1^{(r)} = 0$ otherwise.

Set $\mathcal{N}_U$ is divided into two subsets, i.e., $\mathcal{N}_U = \mathcal{N}_{U,U} \cup \mathcal{N}_{U,O}$, where $\mathcal{N}_{U,U} = \{n \in \mathcal{N}_U \,|\, \sum_{r \in \mathcal{R}} X_n^{(r)}(k)\mathtt{f}_1^{(r)} = 0\}$ is a set of underloaded NFV nodes on which no SFCs traverse other overloaded NFV nodes, and $\mathcal{N}_{U,O} = \{n \in \mathcal{N}_U \,|\, \sum_{r \in \mathcal{R}} X_n^{(r)}(k)\mathtt{f}_1^{(r)} > 0\}$ is a set of underloaded NFV nodes on which at least one SFC traverses other overloaded NFV nodes.

Let $\mathtt{f}_2^{(r)}$ be a binary flag indicating whether SFC $r \in \mathcal{R}$ traverses any NFV nodes in $\mathcal{N}_{U,O}$, with $\mathtt{f}_2^{(r)} = 1$ if $\sum_{n \in \mathcal{N}_{U,O}} X_n^{(r)}(k) > 0$, and $\mathtt{f}_2^{(r)} = 0$ otherwise.

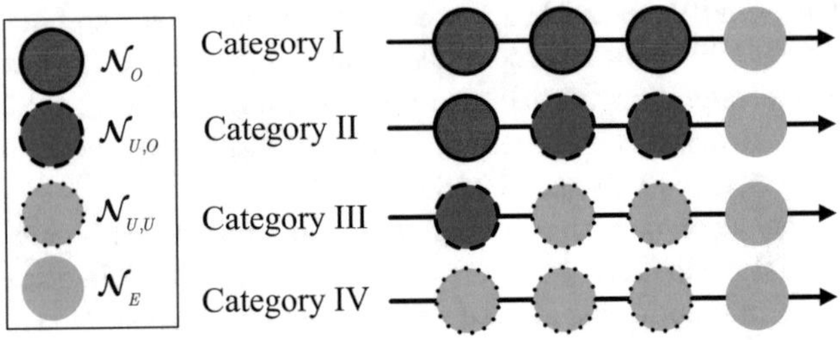

Fig. 3.11 Four SFC categories based on NFV node utilization factors

Accordingly, SFCs are classified into four categories: SFC category I with $\mathtt{f}_1^{(r)} = 1$ and $\mathtt{f}_2^{(r)} = 0$, SFC category II with $\mathtt{f}_1^{(r)} = \mathtt{f}_2^{(r)} = 1$, SFC category III with $\mathtt{f}_1^{(r)} = 0$ and $\mathtt{f}_2^{(r)} = 1$, and SFC category IV with $\mathtt{f}_1^{(r)} = \mathtt{f}_2^{(r)} = 0$, as shown in Fig. 3.11. The distinct features of the four SFC categories are described as follows:

- For an SFC in category I, all the VNFs are placed at NFV nodes with utilization factors no less than threshold η_{th}. There is no opportunity to relax the hop delay bound at any VNF by making the hop delay bound at another VNF more stringent;
- For an SFC in category II, the VNFs are placed at both overloaded and underloaded NFV nodes. Note that the underloaded NFV nodes traversed by SFCs in category II must be from set $\mathcal{N}_{U,O}$. There is an opportunity to relax the hop delay bounds of VNFs placed at the overloaded NFV nodes by shrinking the hop delay bounds of VNFs at the underloaded NFV nodes. With hop delay bound relaxation at the overloaded NFV nodes, the resource utilization at the overloaded NFV nodes can be reduced. If the utilization factor of an overloaded NFV node is reduced to at most η_{th}, the overloaded NFV node is no longer overloaded;
- For an SFC in category III, no VNF is placed at overloaded NFV nodes, with some of the placed NFV nodes from set $\mathcal{N}_{U,U}$ and the remaining from set $\mathcal{N}_{U,O}$. Such an underloaded NFV node $n \in \mathcal{N}_{U,O}$ is traversed by at least one SFC from category II. Hence, if we shrink the hop delay bounds of VNFs placed at the underloaded NFV nodes in $\mathcal{N}_{U,U}$, the hop delay bounds of VNFs at the underloaded NFV nodes in $\mathcal{N}_{U,O}$ can be relaxed, thus releasing more resources for SFC category II sharing resources with SFC category III at underloaded NFV nodes in $\mathcal{N}_{U,O}$. Then, for SFC category II, the hop delay bounds of VNFs at the underloaded NFV nodes in $\mathcal{N}_{U,O}$ can be made more stringent, which allows the hop delay bounds of VNFs placed at the overloaded NFV nodes to be more relaxed. In this way, the utilization factor of the overloaded NFV nodes can be further reduced, and there is more opportunity to turn the overloaded NFV nodes into underloaded;

- For an SFC in category IV, all the VNFs are placed at underloaded NFV nodes in set $\mathcal{N}_{U,U}$ and NFV nodes in set $\mathcal{N}_E$. There is no need to relax any hop delay bound of the VNFs.

Define two sets of delay scaling factors, vertical delay scaling factors $\{\tilde{\alpha}_n(k), \forall n \in \mathcal{N}\}$ and horizontal delay scaling factors $\{\tilde{\beta}^{(r)}(k), \forall r \in \mathcal{R}\}$, with initial values of 1. A two-step delay scaling strategy is used for hop delay bound redistribution, which is described as follows.

3.4.2.1 Step I: Delay Scaling for SFC Category III

Hop delay bounds for SFC category III are relaxed on NFV nodes in $\mathcal{N}_{U,O}$, to release resources for SFC category II, by making hop delay bounds more stringent on NFV nodes in $\mathcal{N}_{U,U}$.

First, for an NFV node $n \in \mathcal{N}_{U,U}$ traversed by SFC category III, the utilization factor is increased to threshold η_{th}, by shrinking hop delay bounds for SFC category III on NFV node $n \in \mathcal{N}_{U,U}$ by a positive factor, $\tilde{\alpha}_n(k)$, less than 1, as derived in Appendix A and given by

$$\tilde{\alpha}_n(k) = \frac{\sum_{(r,h)\in\mathcal{V}} \frac{P_h^{(r)} b_r}{\mathsf{D}^{rh}(k)} \mathrm{x}_n^{rh}(k) \mathtt{f}_2^{(r)}}{[\eta_{th} - \eta_n(k)]\,\mathsf{C}_n + \sum_{(r,h)\in\mathcal{V}} \frac{P_h^{(r)} b_r}{\mathsf{D}^{rh}(k)} \mathrm{x}_n^{rh}(k) \mathtt{f}_2^{(r)}}. \tag{3.27}$$

The preceding delay scaling, called vertical delay scaling, is applied to multiple SFCs belonging to category III on NFV node $n \in \mathcal{N}_{U,U}$. For example, if VNF $V_h^{(r)}$ belongs to SFC category III and is placed at the NFV node $n \in \mathcal{N}_{U,U}$ with vertical delay scaling, the hop delay bound of VNF $V_h^{(r)}$ is scaled by a ratio of $\tilde{\alpha}_n(k)$, i.e., $D^{rh}(k)$ is updated as $\tilde{\alpha}_n(k) D^{rh}(k)$.

Then, hop delay bounds for SFC r in category III on NFV nodes in $\mathcal{N}_{U,O}$ are relaxed by a factor, $\tilde{\beta}^{(r)}(k)$, larger than 1, given by

$$\tilde{\beta}^{(r)}(k) = \frac{\mathsf{D}_r - \sum_{h\in\mathcal{H}_r} \left(\mathsf{D}^{rh}(k) \sum_{n\in\mathcal{N}_{U,U}\cup\mathcal{N}_E} \mathrm{x}_n^{rh}(k) \right)}{\mathsf{D}_r - \sum_{h\in\mathcal{H}_r} \left(\mathsf{D}_p^{rh}(k) \sum_{n\in\mathcal{N}_{U,U}\cup\mathcal{N}_E} \mathrm{x}_n^{rh}(k) \right)} \tag{3.28}$$

where $\mathsf{D}_p^{rh}(k)$ is the old value of $\mathsf{D}^{rh}(k)$ before vertical delay scaling. The preceding delay scaling, called horizontal delay scaling, is applied to multiple hops in an SFC affected by vertical delay scaling. For example, if VNF $V_h^{(r)}$ belongs to SFC category III and is placed at the NFV node $n \in \mathcal{N}_{U,O}$, the hop delay bound of VNF $V_h^{(r)}$ is scaled by a ratio of $\tilde{\beta}^{(r)}(k)$, i.e., $D^{rh}(k)$ is updated as $\tilde{\beta}^{(r)}(k) D^{rh}(k)$.

With the updated hop delay bounds, $\{D^{rh}(k)\}$, the NFV node utilization factors, $\{\eta_n(k)\}$, are updated based on (3.25).

3.4.2.2 Step II: Delay Scaling for SFC Category II

More resources available at NFV nodes in $\mathcal{N}_{U,O}$ from Step I are allocated to SFC category II. First, vertical delay scaling is applied to NFV nodes in $\mathcal{N}_{U,O}$ to increase their utilization factors to η_{th}, through scaling hop delay bounds for SFC category II on it by a vertical delay scaling factor. Then, horizontal delay scaling is applied to SFC category II, by relaxing hop delay bounds for SFC r in category II on NFV nodes in $\mathcal{N}_O$ by a horizontal scaling factor. The scaling factors are similar to that in Step I, given by

$$\tilde{\alpha}_n(k) = \frac{\sum_{(r,h)\in\mathcal{V}} \frac{P_h^{(r)} b_r}{\mathsf{D}^{rh}(k)} \mathrm{x}_n^{rh}(k)\, \mathtt{f}_1^{(r)}}{[\eta_{th} - \eta_n(k)]\, \mathsf{C}_n + \sum_{(r,h)\in\mathcal{V}} \frac{P_h^{(r)} b_r}{\mathsf{D}^{rh}(k)} \mathrm{x}_n^{rh}(k)\, \mathtt{f}_1^{(r)}} \tag{3.29}$$

and

$$\tilde{\beta}^{(r)}(k) = \frac{\mathsf{D}_r - \sum_{h\in\mathcal{H}_r} \left(\mathsf{D}^{rh}(k) \sum_{n\in\mathcal{N}_{U,O}\cup\mathcal{N}_E} \mathrm{x}_n^{rh}(k)\right)}{\mathsf{D}_r - \sum_{h\in\mathcal{H}_r} \left(D_p^{rh}(k) \sum_{n\in\mathcal{N}_{U,O}\cup\mathcal{N}_E} \mathrm{x}_n^{rh}(k)\right)}. \tag{3.30}$$

After the delay scaling procedures in Step II, the hop delay bounds, $\{D^{rh}(k)\}$, and the NFV node utilization factors, $\{\eta_n(k)\}$, are updated, with which the NFV node classification and SFC categories are updated. Specifically, the NFV nodes with vertical delay scaling all have a utilization factor equal η_{th}. The NFV nodes with horizontally delay scaled VNFs have a reduced utilization factor.

Based on the new NFV node loading status, if there remains at least one SFC in category II, i.e., $\mathcal{N}_{U,O} \neq \varnothing$ and $\sum_{r\in\mathcal{R}} \mathtt{f}_1^{(r)} > 0$, it is possible to further reduce the number of overloaded NFV nodes through Step II. Therefore, Step II is performed iteratively until the condition is violated. The iterative delay scaling procedure for a given NFV node utilization factor η_{th} composites one round of hop delay bound redistribution, with the pseudo code given in Algorithm 3.2. The outputs of Algorithm 3.2 are shown in Line 14, where $\mathcal{N}_{O,1} = \{n \in \mathcal{N}_O | \sum_{r\in\mathcal{R}} X_n^{(r)}(k) = 1\}$ denotes a set of overloaded NFV nodes traversed by a single SFC. As shown in the flowchart in Fig. 3.10, a redistribution of hop delay bounds is performed after initialization to check whether migration is required.

3.4.3 *Loop 1: Sequential Migration Decision*

Let binary variable, v, indicate whether migration is required to overcome resource overloading. It is set as 1 initially. After the initial hop delay bound redistribution with $\eta_{th} = \eta_U$, v is updated based on outputs of Algorithm 3.2:

Algorithm 3.2: Redistribution of hop delay bounds

Input: η_{th}, $\{\mathrm{x}_n^{rh}(k)\}$, $\{\mathsf{D}^{rh}(k)\}$
Calculate $\{\eta_n(k)\}$, $\mathcal{N}_O$, $\mathcal{N}_{U,U}$, $\mathcal{N}_{U,O}$, $\mathcal{N}_E$, $\{\mathtt{f}_1^{(r)}\}$, $\{\mathtt{f}_2^{(r)}\}$.
while $\mathcal{N}_{U,O} \neq \varnothing$ *and* $\sum_{r\in\mathcal{R}} \mathtt{f}_1^{(r)} > 0$ **do**
  $\{\tilde{\alpha}_n(k)\} = 1$; $\{\tilde{\beta}^{(r)}(k)\} = 1$.
  if *in the first while loop,* **then**
    Vertical delay scaling at NFV nodes in $\mathcal{N}_{U,U}$.
    Horizontal delay scaling for SFC category III at NFV nodes in $\mathcal{N}_{U,O}$.
    Update $\{\eta_n(k)\}$.
  Vertical delay scaling at NFV nodes in $\mathcal{N}_{U,O}$.
  Horizontal delay scaling for SFC category II at NFV nodes in $\mathcal{N}_O$.
  Update $\{\eta_n(k)\}$, $\mathcal{N}_O$, $\mathcal{N}_{U,U}$, $\mathcal{N}_{U,O}$, $\mathcal{N}_E$, $\{\mathtt{f}_1^{(r)}\}$, $\{\mathtt{f}_2^{(r)}\}$.
Output: $\{\mathsf{D}^{rh}(k)\}$, $\{\eta_n(k)\}$, $\mathcal{N}_O$, $\mathcal{N}_{O,1}$.

- If there are no overloaded NFV nodes with $\eta_n(k) > \eta_{th} = \eta_U$, i.e., $\mathcal{N}_O = \varnothing$, no migration is required, and the value of v is reset to 0;
- If there are still overloaded NFV nodes with $\eta_n(k) > \eta_{th} = \eta_U$, i.e., $\mathcal{N}_O \neq \varnothing$, delay scaling is not sufficient to deal with resource overloading on NFV nodes, but at least one migration is required. Then, v remains as 1, and η_{th} remains as η_U to check whether more migrations are required after a migration decision is made.

When one migration is required, a migration decision procedure is to select one bottleneck NFV node, one SFC to migrate, and one target NFV node. Migration decisions are made greedily to reduce the migration cost. First, a candidate bottleneck NFV node set, $\mathcal{B}$, with $|\mathcal{B}| = |\mathcal{R}|$, is determined. For an SFC, the traversed NFV node with largest hop delay bound is selected as a candidate bottleneck. Then, bottleneck NFV node, n_b, is determined in three cases. In the first case with $\mathcal{B} \cap \mathcal{N}_O \neq \varnothing$, an NFV node in $\mathcal{B} \cap \mathcal{N}_O$ with the largest number of SFCs is selected, given by

$$n_b = \underset{n\in\mathcal{B}\cap\mathcal{N}_O}{\operatorname{argmax}} \sum_{r\in\mathcal{R}} X_n^{(r)}(k). \tag{3.31}$$

In the second case with $\mathcal{B} \cap \mathcal{N}_O = \varnothing$ and $\mathcal{N}_O \backslash \mathcal{N}_{O,1} \neq \varnothing$, n_b is an NFV node in $\mathcal{N}_O \backslash \mathcal{N}_{O,1}$ with the largest utilization factor,

$$n_b = \underset{n\in\mathcal{N}_O\backslash\mathcal{N}_{O,1}}{\operatorname{argmax}} \eta_n(k). \tag{3.32}$$

In the third case with $\mathcal{B} \cap \mathcal{N}_O = \mathcal{N}_O \backslash \mathcal{N}_{O,1} = \varnothing$, an NFV node in $\mathcal{B}$ whose SFCs traverse the largest number of overloaded NFV nodes is selected, given by

Algorithm 3.3: Heuristic algorithm for the problem in (3.18)

1 **Input:** Step size $\eta_{\Delta,0}$, precision η^{ϵ}_{Δ}
2 **Initialize:** $\{\mathrm{x}_n^{rh}(k)\}$, $\{\mathsf{D}^{rh}(k)\}$
3 **Let:** $\mathrm{v} = 1$, $\eta_{th} = \eta_U$, $\eta_\Delta = \eta_{\Delta,0}$
4 **while** $\mathrm{v} == 1$ **do**
5   Redistribution of hop delay bounds: Update $\{\mathsf{D}^{rh}(k)\}$, $\{\eta_n(k)\}$, $\mathcal{N}_O$, $\mathcal{N}_{O,1}$ according to Algorithm 3.2;
6   **if** $\mathcal{N}_O \neq \varnothing$, **then**
7   Update $\{\mathrm{x}_n^{rh}(k)\}$ according to the migration decision procedure;
8   **else**
9         $\mathrm{v} = 0$;
10 **while** $\eta_\Delta > \eta^{\epsilon}_{\Delta}$ **do**
11   **if** $\mathcal{N}_O = \varnothing$, **then**
12   **if** $\eta_\Delta \neq \eta_{\Delta,0}$, **then** $\eta_\Delta = \eta_\Delta/2$;
13         $\eta_{th} = \eta_{th} - \eta_\Delta$;
14   **else**
15         $\eta_\Delta = \eta_\Delta/2$;
16         $\eta_{th} = \eta_{th} + \eta_\Delta$;
17   Redistribution of hop delay bounds: Update $\{\mathsf{D}^{rh}(k)\}$, $\{\eta_n(k)\}$, $\mathcal{N}_O$, $\mathcal{N}_{O,1}$ according to Algorithm 3.2.
18 **Output:** Sub-optimal solution to the problem in (3.18).

$$n_b = \operatorname*{argmax}_{n \in \mathcal{B}} \sum_{r \in \mathcal{R}} \left(X_n^{(r)}(k) \sum_{n' \in \mathcal{N}_O} X_{n'}^{(r)}(k) \right). \tag{3.33}$$

Next, an SFC to migrate from n_b and a target NFV node to accommodate the migrated SFC are jointly selected to minimize the migration cost, i.e., the weighted sum of normalized transmission resource overhead for state transfer and number of new virtual links for flow rerouting. In this way, ω_2 and ω_3 are considered in the heuristic algorithm. If there are multiple choices, an SFC with the largest resource demand is migrated to the closest target NFV node.

After a migration decision, redistribution of hop delay bounds is performed again to check whether more migrations are required. Migration decisions are performed sequentially, each followed by a redistribution of hop delay bounds, until no more migrations are required, and v is reset to 0. The alternate resource scaling and migration in Loop 1 is summarized in Lines 4–9 of Algorithm 3.3.

3.4.4 Loop 2: Iterative Resource Utilization Threshold Update

Once Loop 1 is ended with $\mathrm{v} = 0$, no more migrations are required, and all the NFV nodes have a utilization factor less than or equal to the initial threshold $\eta_{th} = \eta_U$.

The gap between the smallest and largest NFV node utilization factors can be large, which is undesired in terms of load balancing. Actually, if one SFC traverses two NFV nodes with different utilization factors, its hop delay bound can be relaxed at the NFV node with the larger utilization factor and be shrunk on the other NFV node, to reduce the gap between the two utilization factors. Based on this idea, an iterative procedure with alternate threshold updating and redistribution of hop delay bounds is performed to gradually reduce the gap and to balance the NFV node utilization factors.

Let η_Δ be a step size to update η_{th}, with initial value $\eta_{\Delta,0}$ and being updated before each η_{th} update. The threshold, η_{th}, is first reduced stepwise from the initial value η_U, with an initial step size $\eta_\Delta = \eta_{\Delta,0}$, as indicated by Line 13 in Algorithm 3.3. For each threshold update, a redistribution of hop delay bounds is performed. This process is repeated until some overloaded NFV nodes are detected after hop delay bound redistribution, i.e., $\mathcal{N}_O \neq \varnothing$.

The emergence of overloaded NFV nodes indicates that the latest step of threshold reduction is too aggressive. Then, a binary search between the latest two threshold values is performed. First, the step size is reduced by half, i.e., $\eta_\Delta = \eta_\Delta/2$, and threshold η_{th} is increased by a step, as indicated by Lines 15–16 of Algorithm 3.3. Depending on the results of hop delay bound redistribution in Line 17 of Algorithm 3.3, threshold η_{th} is increased or decreased by a step, with a step size reduced by half for each iteration, until a sufficient precision is reached. A predefined precision, η_Δ^ϵ, for η_Δ is set as the stop condition.

3.4.5 Complexity Analysis

We first analyze the time complexity of Algorithm 3.2. Delay scaling for SFC category III is performed once, using at most $O(\sum_{n\in\mathcal{N}} |\mathcal{V}|)$ time. Delay scaling for SFC category II is performed iteratively until there are no new NFV nodes in $\mathcal{N}_{U,O}$ or there are no overloaded SFCs. The worst case happens when each round of delay scaling for SFC category II transforms a single NFV node in $\mathcal{N}_O$ to a new NFV node in $\mathcal{N}_{U,O}$, consuming $O(\sum_{n\in\mathcal{N}_O} \sum_{n\in\mathcal{N}} |\mathcal{V}|)$ time. Thus, the complexity of Algorithm 3.2 is $O(\sum_{n\in\mathcal{N}_O} \sum_{n\in\mathcal{N}} |\mathcal{V}|)$, upper bounded by $O(|\mathcal{N}|^2|\mathcal{V}|)$. The complexity of the migration decision procedure is dominated by the selection of SFC to migrate in the third case, which requires a running time of $O(|\mathcal{N}||\mathcal{R}|^2)$. In Algorithm 3.3, at most $|\mathcal{V}|$ sequential migration decisions are performed followed by $[\frac{1}{\eta_{\Delta,0}} + \log_2(\frac{\eta_{\Delta,0}}{\eta_\Delta^\epsilon})]$ iterations of threshold updating. In each iteration, hop delay bounds are readjusted. Therefore, the worst case running time of Algorithm 3.3 is $|\mathcal{V}|[O(|\mathcal{N}||\mathcal{R}|^2)+O(|\mathcal{N}|^2|\mathcal{V}|)]+[\frac{1}{\eta_{\Delta,0}}+\log_2(\frac{\eta_{\Delta,0}}{\eta_\Delta^\epsilon})]O(|\mathcal{N}|^2|\mathcal{V}|)$, which is simplified to $O(|\mathcal{N}|^2|\mathcal{V}|^2)$ when $|\mathcal{R}|^2 < |\mathcal{N}||\mathcal{V}|$.

3.5 Simulation Results

In this section, simulation results are presented to evaluate the MIQCP and heuristic solutions for the delay-aware flow migration problem. Two time intervals are considered: $(k-1)$ and k, representing the current and next time intervals respectively. We use two mesh networks with 64 NFV nodes and 256 NFV nodes to represent the virtual resource pool. Virtual links exist only between neighboring NFV nodes.

In the 64-node network, we consider known SFC mapping for time interval $(k-1)$, with three SFCs initially mapped to the virtual resource pool. Specifically, SFC 3 (in blue color) shares two NFV nodes with SFC 1 (in yellow color) and one of them also with SFC 2 (in red color), as illustrated in Fig. 3.12. The sources and destinations of SFCs are omitted for clarity. In the 256-node network, we consider different numbers of SFCs, with [2, 5] VNFs in each one, randomly distributed in

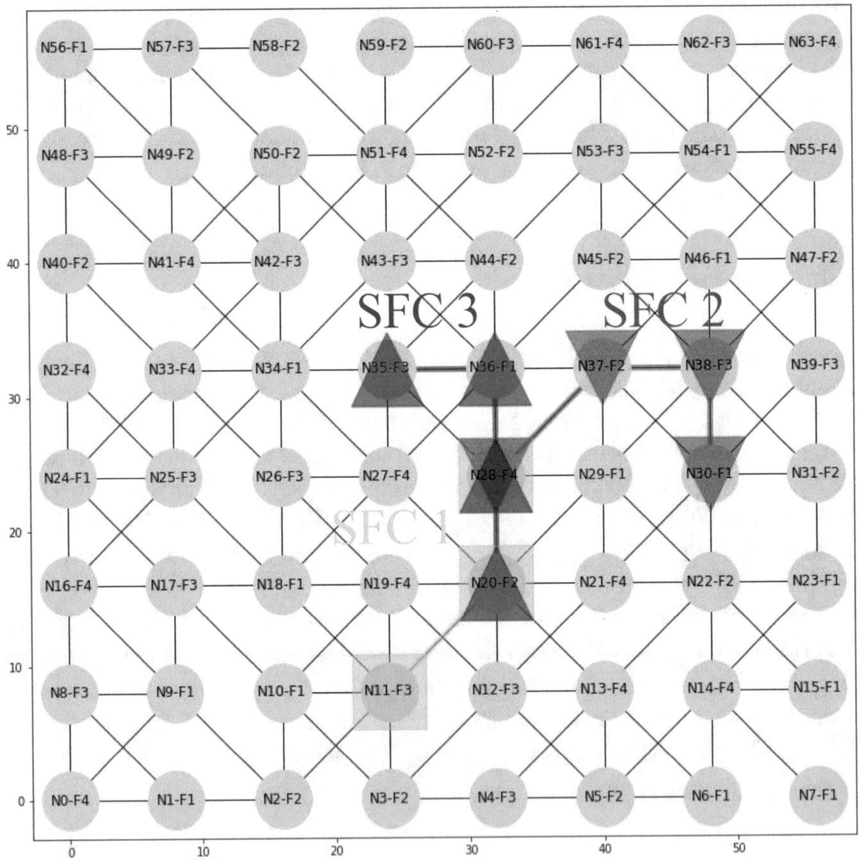

Fig. 3.12 Initial SFC mapping in a 64-node mesh network with 3 SFCs

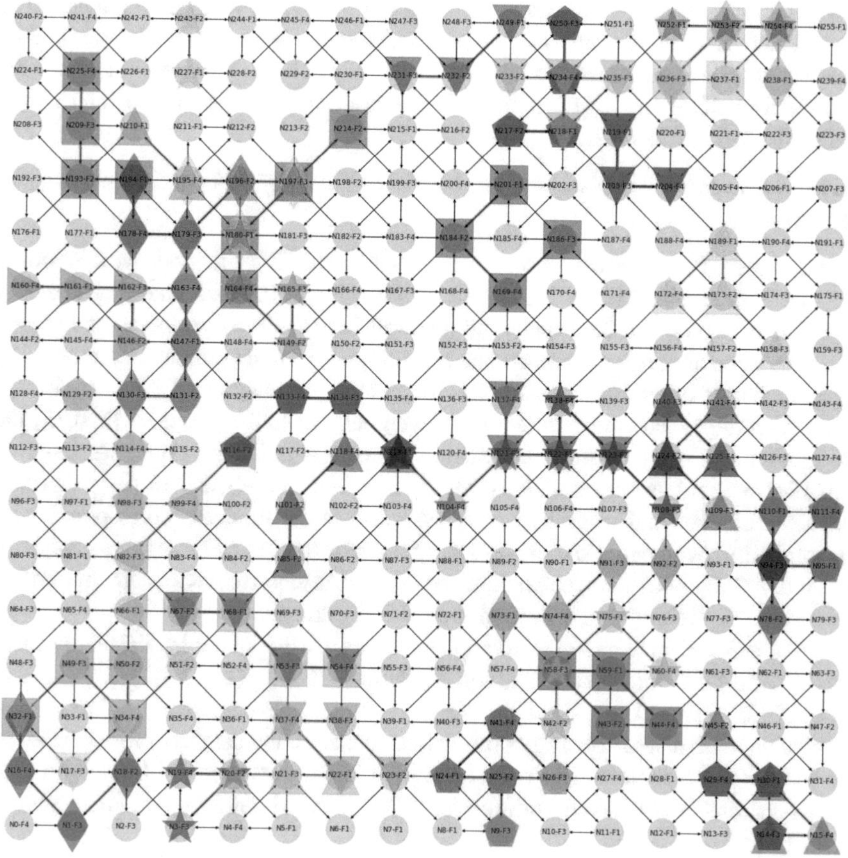

Fig. 3.13 Initial SFC mapping in a 256-node mesh network with 45 SFCs

the network during time interval $(k-1)$. An example of random SFC distribution with 45 SFCs in the 256-node network is illustrated in Fig. 3.13. All SFCs in both networks have an initial traffic rate of 200 packet/s during the previous time interval $(k-1)$, i.e., $\lambda^{(r)}(k-1) = 200, \forall r \in \mathcal{R}$.

We set a ratio of 0.01 between the switching time and the CPU polling period. The upper bound, η_U, for $\eta(k)$, is 0.95. We consider same processing density for all VNFs and a maximum NFV node processing rate of 1000 packet/s. The average E2E delay requirement and the maximal tolerable service downtime for each SFC are 20 ms and 5 ms, respectively. For VNF states, the size is a constant, equal to 10 *bytes*, thus requiring at least a transmission rate B_{min} of 16 *kbit/s* for a state transfer. Under the simulation setup, the total normalized transmission resource overhead for state transfer is equal to the total number of migrations. For the weights, we set $\omega_2 = 2\omega_3$. Then, $\omega_1 < \frac{2}{3\eta_U+2} = 0.4123$ should be satisfied to penalize migration more than imbalanced loading. In this case, 0.4123 is the worst-case boundary for

ω_1, to guarantee the penalization preference if ω_1 is less than the boundary. The initial step size and the precision for threshold updating are set to 0.1 and 0.0001, respectively. We implement both the MIQCP and heuristic solutions in python. We use NetworkX to simulate the network scenario, and Gurobi python interface to solve the MIQCP problem.

Load Balancing and Reconfiguration Overhead Trade-off with the Optimal MIQCP Solution

We use the 64-node virtual resource pool with three SFCs to evaluate the performance of the optimal MIQCP solution with varying traffic load under three sets of weights in (3.17), and investigate the trade-off between load balancing and migration cost. Performance metrics are the maximum NFV node utilization factor, $\eta(k)$, the number of migrations, $N_m(k)$, and the number of new virtual links, $N_e(k)$, for flow rerouting. We explore three sets of weights. For $\{\omega_1, \omega_2, \omega_3\} = \{1, 0, 0\}$, the migration cost is not optimized but load balancing is the focus, corresponding to a load balancing flow migration (LBFM) strategy. For $\{\omega_1, \omega_2, \omega_3\} = \{0, \frac{2}{3}, \frac{1}{3}\}$, $\eta(k)$ is not optimized but migration cost reduction is emphasized, corresponding to a minimum overhead flow migration (MOFM) strategy. For $\{\omega_1, \omega_2, \omega_3\} = \{0.4, 0.4, 0.2\}$, both load balancing and migration cost reduction are important, corresponding to a hybrid flow migration (HFM) strategy.

Figure 3.14 show the SFC mapping results in the 64-node virtual resource pool with different flow migration strategies at $\{\lambda^{(1)}(k), \lambda^{(2)}(k), \lambda^{(3)}(k)\} = \{700, 200, 700\}$ packet/s. For the LBFM strategy, SFCs completely separate from each other even at a relatively low traffic load to maximally balance traffic load in the virtual resource pool, as illustrated in Fig. 3.14a. However, only a limited number of migrations are observed with the MOFM and HFM strategies, as illustrated in Fig. 3.14b.

Next, we fix the traffic loads of SFC 1 and SFC 2 at $\lambda^{(1)}(k) = 600$ packet/s and $\lambda^{(2)}(k) = 200$ packet/s, and vary the traffic rate of SFC 3, $\lambda^{(3)}(k)$, from 200 packet/s to 740 packet/s. Beyond 740 packet/s, the optimization problem becomes infeasible due to computing resource constraints and average E2E delay constraints. Figure 3.15 shows performance of three flow migration strategies with the increase of $\lambda^{(3)}(k)$, for $\eta_U = 0.95$.

LBFM Strategy It is observed that the maximum NFV node utilization factor, $\eta(k)$, is dominated by SFC 1 with traffic rate $\lambda^{(1)}(k) = 600$ packet/s, when $\lambda^{(3)}(k)$ is relatively small. Correspondingly, $\eta(k)$ shows a flat trend first, with the increase of $\lambda^{(3)}(k)$ from 200 packet/s to 500 packet/s. With the increase of $\lambda^{(3)}(k)$ beyond 500 packet/s, $\eta(k)$ turns to be dominated by SFC 3. Here, we should note that SFC 1 and SFC 3 have the same E2E delay requirement, but SFC 3 has one more VNF in the chain, resulting in a more stringent average delay requirement at each VNF and a higher resource demand at the placed NFV node. That's why we see the turning point in $\eta(k)$ at $\lambda^{(3)}(k) = 500$ packet/s which is less than the traffic rate of SFC 1. For the migration cost, both the number of VNF migrations, $N_m(k)$, and the number

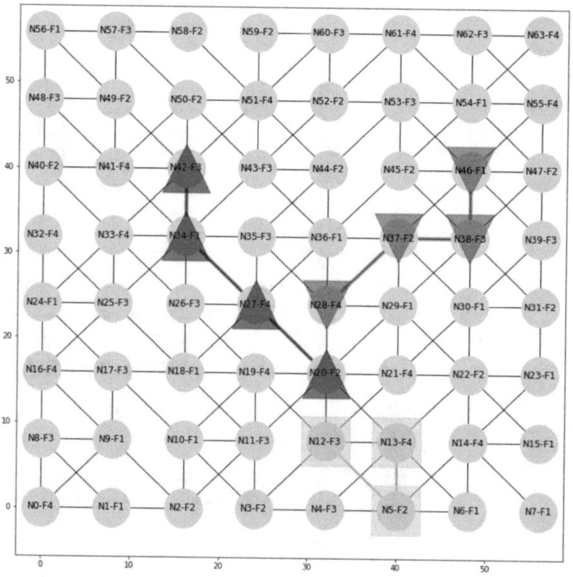

(a)

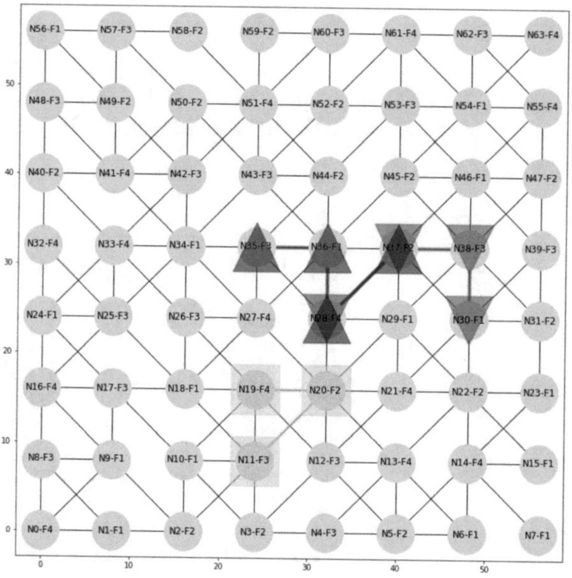

(b)

Fig. 3.14 New SFC mapping in a 64-node mesh network for $\lambda^{(1)}(k) = 700$, $\lambda^{(2)}(k) = 200$, and $\lambda^{(3)}(k) = 700$. (**a**) Free migrations (LBFM). (**b**) Limited migrations (MOFM, HFM)

Fig. 3.15 Performance of three flow migration strategies with respect to $\lambda^{(3)}(k)$. (**a**) LBFM strategy. (**b**) MOFM strategy. (**c**) HFM strategy

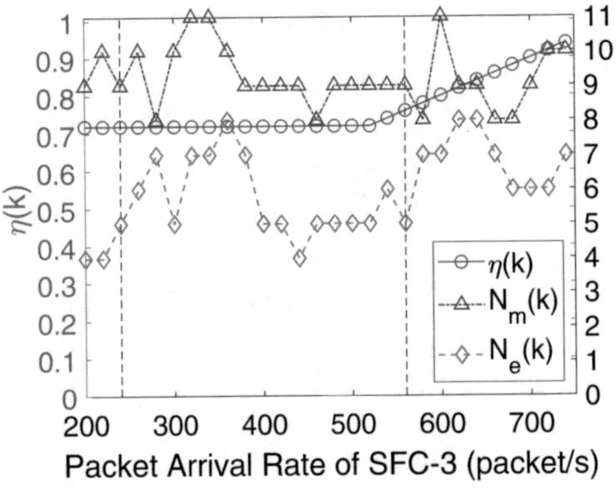

(a)

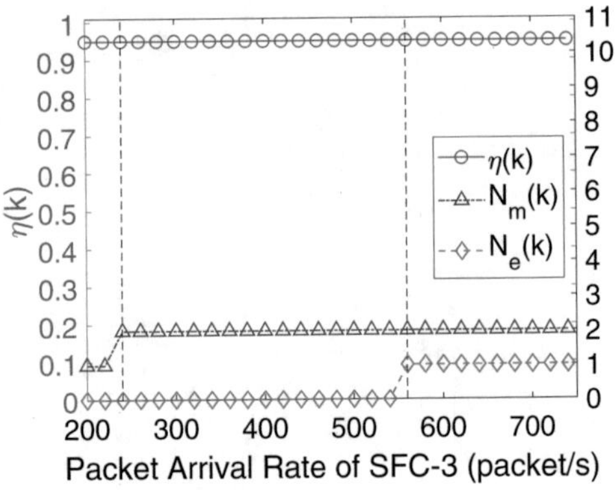

(b)

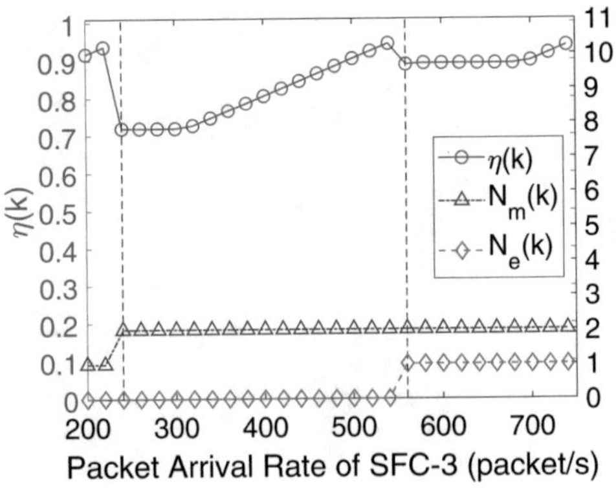

(c)

of new virtual links, $N_e(k)$, are high and vary randomly with the traffic load, since they are not optimized in the LBFM strategy.

MOFM Strategy Both the number of VNF migrations, $N_m(k)$, and the number of new virtual links, $N_e(k)$, show a step-wise increasing trend with the increase of $\lambda^{(3)}(k)$. The MOFM strategy determines the lower bounds of $N_m(k)$ and $N_e(k)$, since it emphasizes migration cost minimization. However, we observe that $\eta(k)$ is fixed at $\eta_U = 0.95$, since it is not optimized in the MOFM strategy.

HFM Strategy A trade-off among the three performance metrics is observed. With the increase of $\lambda^{(3)}(k)$, the maximum NFV node utilization factor, $\eta(k)$, drops sharply when either the number of VNF migrations, $N_m(k)$, or the number of new virtual links, $N_e(k)$, is increased by 1. When $N_m(k)$ and $N_e(k)$ stay stable, $\eta(k)$ shows either a linear increasing or a flat trend. The linear increasing trend indicates the dominance by $\lambda^{(3)}(k)$, while the flat trend indicates the dominance by $\lambda^{(1)}(k)$. The same step-wise increasing trends in the curves of $N_m(k)$ and $N_e(k)$ as the MOFM strategy are observed. Compared with LBFM and MOFM strategies, HFM strategy approaches to the lower bounds of $N_m(k)$ and $N_e(k)$, while keeping $\eta(k)$ at a medium level. The relationships among $\eta(k)$, $N_m(k)$, and $N_e(k)$ demonstrate the trade-off between load balancing and migration cost.

Next, we focus on the HFM strategy and examines its performance. Figure 3.16 shows the performance of HFM strategy with both $\lambda^{(1)}(k)$ and $\lambda^{(3)}(k)$ increased from 200 packet/s to 740 packet/s and $\lambda^{(2)}(k)$ fixed at 200 packet/s, for $\eta_U = 1$. We see a two-dimensional zigzag trend for $\eta(k)$ and two-dimensional step-wise increasing trends for both $N_m(k)$ and $N_e(k)$, with the increase of $\lambda^{(1)}(k)$ and $\lambda^{(3)}(k)$. For $\lambda^{(1)}(k) = 280, 480, 700$ packet/s, the performance of HFM strategy with the increase of $\lambda^{(3)}(k)$ is shown in Fig. 3.17. For a given value of $\lambda^{(1)}(k)$, the performance is similar to that in Fig. 3.15.Here, we focus on how the performance curves are different with different values of $\lambda^{(1)}(k)$. Since SFC 1 and SFC 3 share common NFV nodes during interval $k - 1$, we see more migrations with a higher traffic load for SFC 1, to satisfy the QoS requirements of both SFCs. For the turning points on the curve of $N_m(k)$, including the turning points from $N_m(k) = 0$ to $N_m(k) = 1$, from $N_m(k) = 1$ to $N_m(k) = 2$, and from $N_m(k) = 2$ to $N_m(k) = 3$, the corresponding value of $\lambda^{(3)}(k)$ (if there is one) is decreased with the increase of $\lambda^{(1)}(k)$. Similar observations are shown on the curves of $N_e(k)$. Although the curves of $\eta(k)$ all show the zigzag trends with different values of $\lambda^{(1)}(k)$, the lower bounds of $\eta(k)$ are increased for larger $\lambda^{(1)}(k)$. The lower bound of $\eta(k)$ depends the overall traffic loading in the virtual resource pool.

Average End-to-End Delay with the Optimal MIQCP Solution

We use the 64-node mesh virtual network topology with three SFCs to evaluate the benefit of flow migration with joint VNF migration and resource scaling on the average E2E delay performance. The optimal MIQCP solution is used to solve the flow migration optimization model. Without flow migration, VNF migrations are not allowed and only local resource scaling is enabled. We compare the average E2E delay of SFCs with and without flow migration when traffic load increases.

Fig. 3.16 Performance of the HFM flow migration strategy with respect to $\lambda^{(1)}(k)$ and $\lambda^{(3)}(k)$. (**a**) $\eta(k)$. (**b**) $N_m(k)$. (**c**) $N_e(k)$

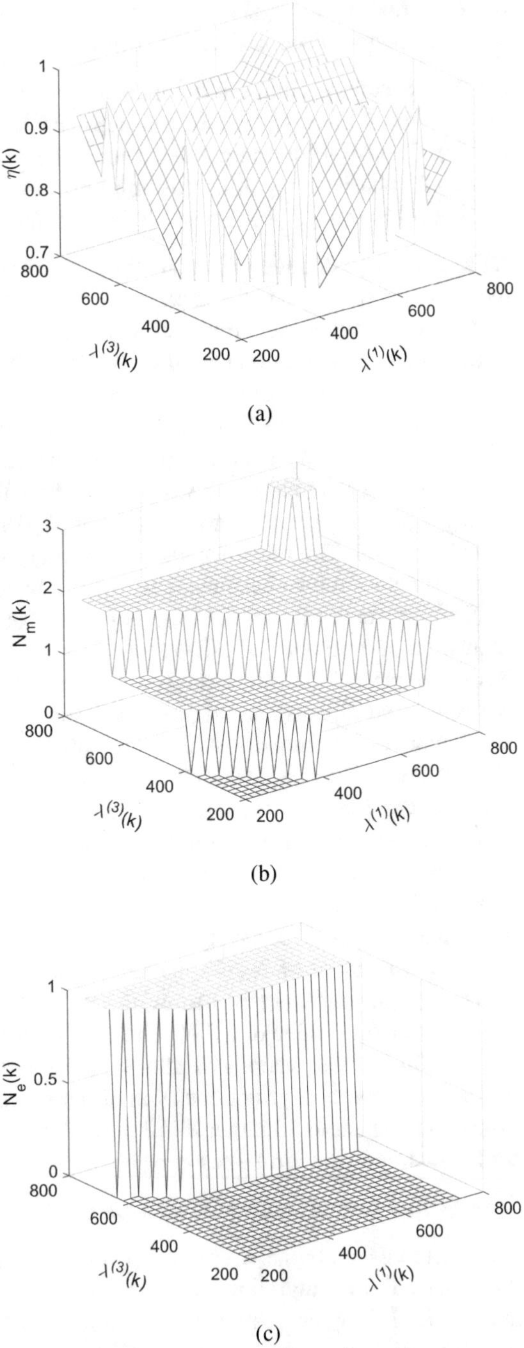

Fig. 3.17 Performance trade-off with the increase of $\lambda^{(3)}(k)$ for (**a**) $\lambda^{(1)}(k) = 280$ packet/s; (**b**) $\lambda^{(1)}(k) = 480$ packet/s; (**c**) $\lambda^{(1)}(k) = 780$ packet/s

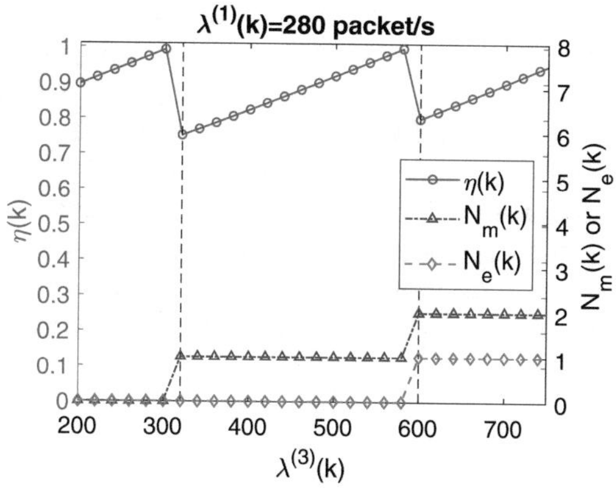

(a)

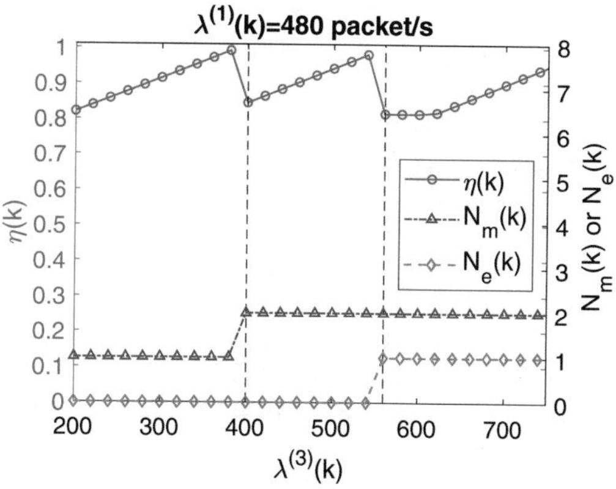

(b)

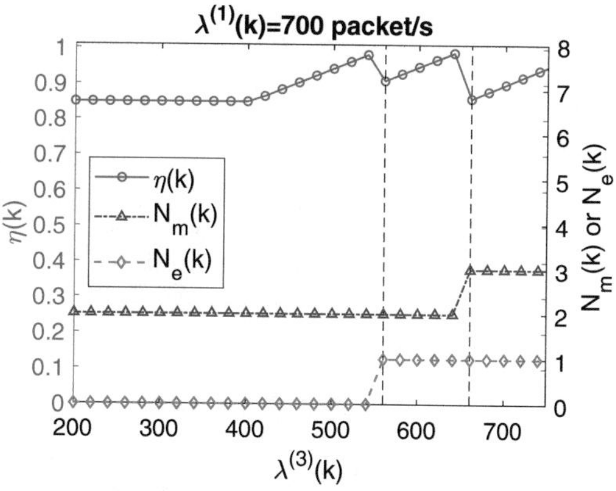

(c)

With the initial traffic rates $\lambda^{(1)}(k-1) = \lambda^{(2)}(k-1) = \lambda^{(3)}(k-1) = 200$ packet/s, we set $\lambda^{(1)}(k)$ and $\lambda^{(2)}(k)$ unchanged at 200 packet/s for SFC 1 and SFC 2, and vary the traffic load of SFC 3, i.e., $\lambda^{(3)}(k)$, to evaluate the E2E delay performance of SFC 3 with traffic variations.

Instead of the hard E2E delay constraints in the original problem, we allow some margins $\boldsymbol{\delta} = \{\delta_r\}$ to the E2E delay requirements $\sum_{h\in\mathcal{H}_r}\sum_{n\in\mathcal{N}} x_n^{rh}(k)d_n^{rh}(k) \le \mathsf{D}_r + \delta_r$, and additionally minimize the aggregate delay margins in an alternative objective function $O(k) + \alpha_4 \sum_{r\in\mathcal{R}} \delta_r$. We confine $\delta_1 = \delta_2 = 0$, and only optimize δ_3. Figure 3.18 shows the average E2E delay of three SFCs without and

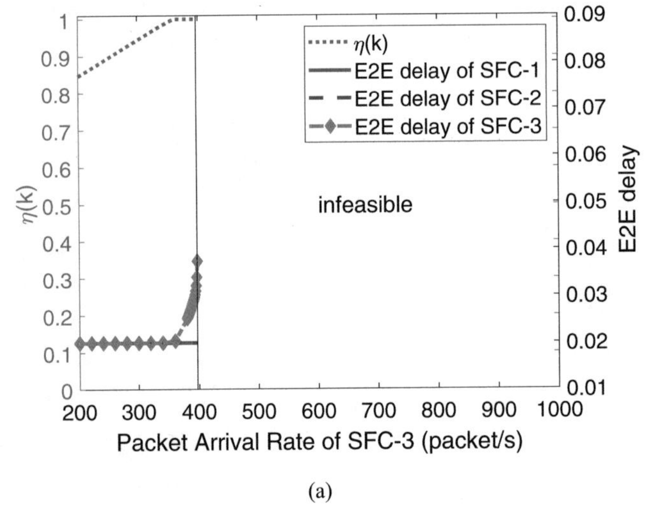

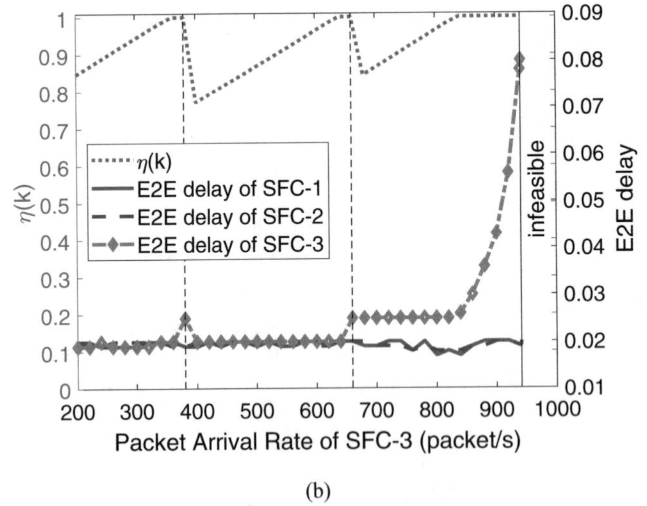

Fig. 3.18 Average E2E delay of SFCs with optimized delay margin. (**a**) Without flow migration. (**b**) With flow migration

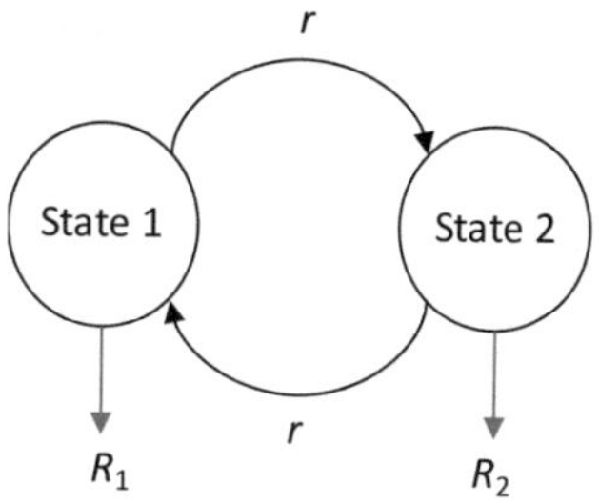

Fig. 3.19 A two-state MMPP traffic model

with flow migration, for $\eta_U = 1$. With the zigzag trend of $\eta(k)$ in Fig. 3.18b, we can see at which traffic rates flow migrations happen. Benefiting from flow migrations, the feasible set of $\lambda^{(3)}(k)$ enlarges from [200, 397] packet/s to [200, 941] packet/s.

End-to-End Delay Performance in Packet-level Simulation

Under the 3-SFC and 64-node network setting, we also carry out packet-level simulations using network simulator OMNeT++ to evaluate the E2E delay performance with and without flow migration. Joint VNF migration and resource scaling is performed based on Poisson traffic model with target rate $\lambda^{(3)}(k)$, if flow migration is allowed. Otherwise, only local computing resource scaling is performed based on Poisson traffic model with target rate $\lambda^{(3)}(k)$.

In the packet-level simulations, the packet arrivals for both SFC 1 and SFC 2 are Poisson. For SFC 3, not only Poisson but also Markov-Modulated Poisson Process (MMPP) packet arrivals are simulated, to verify the effectiveness and accuracy of our flow migration model in the presence of traffic burstiness. In terms of the MMPP packet arrivals, we use a two-state MMPP model with a same transition rate between the two states and an average traffic rate of $\lambda^{(3)}(k) = \frac{R_1+R_2}{2}$ packet/s, with R_1 and R_2 being the individual average packet arrival rates at the two states, as shown in Fig. 3.19. A larger gap between R_1 and R_2 indicates a higher level of traffic burstiness.

For SFC 3, we use "Poisson-$\lambda^{(3)}(k)$" to represent a Poisson traffic model with average packet arrival rate of $\lambda^{(3)}(k)$, and "Poisson" to represent a group of Poisson traffic models with different values of $\lambda^{(3)}(k)$. For the MMPP traffic arrivals, we use "MMPP-R_1-R_2" to represent an MMPP traffic model with state-dependent average packet arrival rates, R_1 and R_2, with $R_1 + R_2 = 2\lambda^{(3)}(k)$. Here, we only consider MMPP traffic models with $R_1, R_2 \geq 50$ packet/s. We use "MMPP-q_1-q_2" to represent a group of MMPP traffic models, where q_1 and q_2 are ratios between state-dependent rates and the average traffic rate, with $q_1 + q_2 = 2$. For example, for the group of "MMPP-1.6-0.4" traffic models, the state-dependent rates are $R_1 = q_1\lambda^{(3)}(k)$ and $R_2 = q_2\lambda^{(3)}(k)$. When $\lambda^{(3)}(k) = 500$ packet/s, the state-dependent rates are 800 packet/s and 200 packet/s respectively.

In each simulation, under the new resource allocation for a target traffic load, which is given by the joint migration and resource scaling decisions with flow migration or given by the local resource scaling decisions without flow migration,

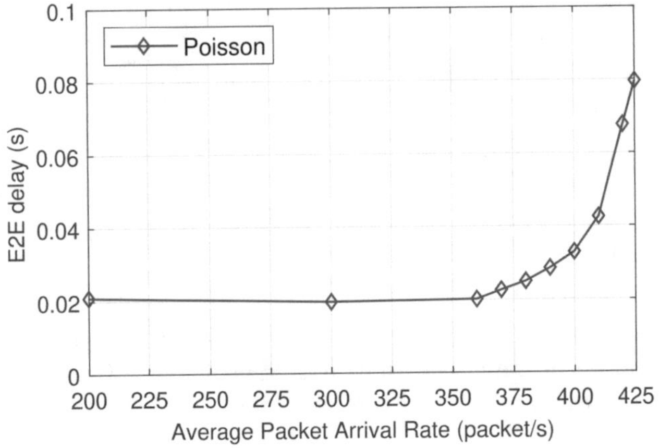

Fig. 3.20 Average E2E delay without flow migration

SFC 3 has packet arrivals in either a Poisson traffic model or an MMPP traffic model. The network simulation time is 5000 s, to collect sufficient packet delay information of SFC 3 under the new resource allocation plan, for a smooth characterization of the E2E delay distribution and an accurate estimation the average E2E delay.

Figure 3.20 shows the average E2E delay of SFC 3 for Poisson traffic arrivals at different target traffic loads without flow migration. A group of Poisson traffic models are evaluated, with increasing rate from 200 packet/s to 425 packet/s. A flat trend is observed in the average E2E delay, with the increase of Poisson traffic rate from 200 packet/s to 360 packet/s, which is followed by an exponential increasing trend, with the further increase of Poisson traffic rate beyond 360 packet/s. The flat trend corresponds to feasible traffic rates for E2E delay guarantee with local computing resource scaling. Beyond 360 packet/s, local resources are not sufficient, resulting in an exponential increase of E2E delay.

Figure 3.21 shows the average E2E delay of SFC 3 for both Poisson and MMPP traffic arrivals at different target traffic loads with flow migration. A group of Poisson traffic models are evaluated, with increasing rate from 200 packet/s to 700 packet/s. Three groups of MMPP traffic models are evaluated, with (1) $q_1 = 1.2$ and $q_2 = 0.8$, (2) $q_1 = 1.4$ and $q_2 = 0.6$, and (3) $q_1 = 1.6$ and $q_2 = 0.4$, respectively. With joint VNF migration and computing resource scaling, we observe that the E2E delay requirement is satisfied for all the Poisson traffic models with rate in [200, 700] packet/s. Without flow migration, the VNFs sharing a common NFV node gradually overload the NFV node with traffic load increase, thus the E2E delay requirement cannot be satisfied at a relatively low traffic load. However, with flow migration, the VNFs can migrate to alternative NFV nodes with sufficient resources, thus accommodating more traffic from the SFCs.

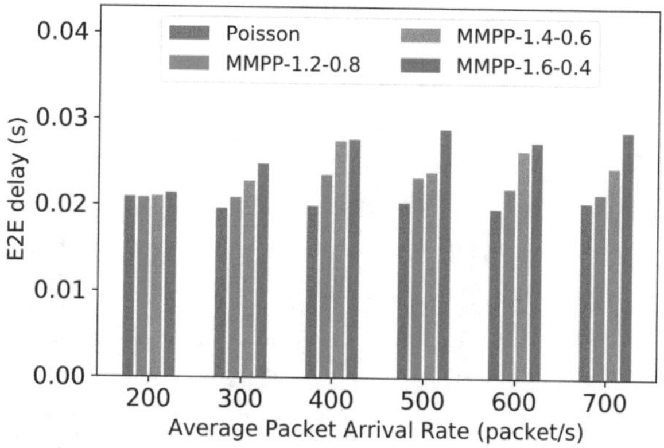

Fig. 3.21 Average E2E delay comparison with flow migration

Table 3.1 E2E delay parameters for Poisson and MMPP traffic models with flow migration

Traffic models	Average E2E delay	P_{D_3}	P_{2*D_3}	P_{3*D_3}
Poisson-300	0.0189 s	66.67%	93.43%	99.22%
MMPP-350-250	0.0206 s	65.24%	91.76%	97.81%
MMPP-450-150	0.0227 s	60.36%	87.87%	96.76%
MMPP-500-100	0.0233 s	58.21%	85.92%	96.67%
MMPP-550-50	0.0257 s	54.08%	82.32%	93.15%
Poisson-322	0.0264 s	49.83%	84.15%	95.22%

At a certain average traffic rate $\lambda^{(3)}(k)$, the average E2E delay degrades with more traffic burstiness. For example, the "MMPP-1.6-0.4" traffic models perform worse than the "MMPP-1.4-0.6" traffic models. However, even the "MMPP-1.6-0.4" traffic models for average rate in [400, 700] packet/s with flow migration performs much better than the "Poisson" traffic models for average rate larger than 360 packet/s without flow migration, indicating that the delay performance gain from flow migration is not significantly compromised by the traffic burstiness.

Figures 3.20 and 3.21 demonstrate the benefit of flow migration in terms of average E2E delay guarantee with traffic load increase, and the performance impact due to traffic burstiness. Next, we examine the E2E delay performance of SFC 3 with flow migration in more details. Specifically, the E2E delay distribution is characterized based on the per-packet E2E delay measurements in the simulations. Both Poisson and MMPP traffic models are evaluated.

Figure 3.22 shows the probability density function (PDF) and cumulative distribution function (CDF) of the E2E delay for SFC 3 with different traffic models, including "Poisson-300", "MMPP-350-250", "MMPP-450-150", "MMPP-500-100", "MMPP-550-50", and "Poisson-322". All the MMPP traffic models have the same average traffic rate of 300 packet/s as "Poisson-300". The flow migration optimization problem is solved based on a target traffic load of $\lambda^{(3)}(k) =$ 300 packet/s. For traffic models with an average traffic rate of 300 packet/s, the E2E delay performance degrades with more traffic burstiness indicated by a

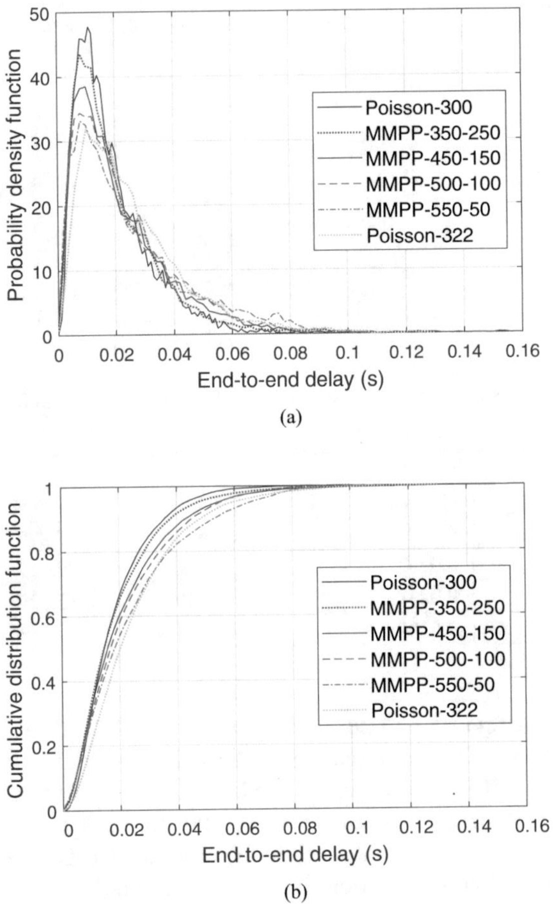

Fig. 3.22 E2E delay distribution for Poisson and MMPP traffic models with flow migration. (**a**) Probability distribution function (PDF). (**b**) Cumulative function distribution (CDF)

larger difference between R_1 and R_2. With more traffic burstiness, the E2E delay distribution has a longer tail. Define P_B as the percentage of packets experiencing an E2E delay within a required delay bound B. Some important parameters including the average E2E delay, P_{D_3}, $P_{2*\mathrm{D}_3}$, and $P_{3*\mathrm{D}_3}$ are summarized in Table 3.1. With more traffic burstiness at a certain average traffic rate, the E2E delay performance degrades on each parameter. However, all MMPP traffic models perform better on most parameters than "Poisson-322", a Poisson traffic model with a slightly larger traffic rate. It demonstrates that, under the flow migration plan for target traffic load $\lambda^{(3)}(k) = 300$ packet/s, the performance degradation with the MMPP traffic models is less significant compared with that of the "Poisson-322" traffic model, indicating that our flow migration model can accommodate some traffic burstiness without a significant degradation on the E2E delay performance.

However, we observe that even for the "Poisson-300" traffic model, at most 66.67% packets experience an E2E delay within D_r, which is not a surprising result since only the average E2E delay performance is guaranteed in our flow migration model. This observation inspires our research in Chaps. 4 and 5, which investigates how to guarantee E2E delay within bound in probability.

In order to increase the ratio of packets within delay bound D_r, i.e., guarantee a larger P_{D_3}, more resources should be allocated than that determined by the flow migration optimization model. We examine the E2E delay performance of SFCs with Poisson traffic arrivals of lower rates than the target rate. Specifically, we compare the performance of the "Poisson-200" and "Poisson-250" traffic models with that of the "Poisson-300" traffic models, under the flow migration plan for a target traffic load of 300 packet/s. Results are shown in Fig. 3.23 and Table 3.2. We observe an improvement on the E2E delay performance with lower traffic rates. It can be inferred that, if we want to guarantee a larger P_{D_3}, either data rate should be reduced or more resources should be allocated.

Comparison Between MIQCP and Heuristic Solutions

Under the 64-node network setup with three SFCs, we compare the MIQCP and heuristic solutions in terms of their cost sensitivity to different weights in (3.17). With $\omega_2 = 2\omega_3$ and $\omega_1 + \omega_2 + \omega_3 = 1$, three cost metrics including the maximum NFV node utilization factor, $\eta(k)$, the migration cost, $2N_m(k) + N_e(k)$, and the total cost, $\omega_1\eta(k) + (1 - \omega_1)(2N_m(k) + N_e(k))$, are evaluated. The first two costs are partial costs. Although the heuristic solution is in principle insensitive to ω_1, we use the same definition of total cost for a fair comparison.

The three cost metrics for both the MIQCP and heuristic solutions with respect to ω_1 are shown in Fig. 3.24. In both solutions, the total cost approaches the migration cost, for ω_1 close to 0, and approaches the maximum NFV node utilization factor, for ω_1 close to 1. We see constant partial costs with respect to ω_1 for the heuristic solution, which is consistent with the design principle. Both partial costs for MIQCP solution show a stable trend for small and medium values of ω_1 in a range larger than the theoretical worst-case range $(0, 0.4123)$. For large values of ω_1, the migration cost of the MIQCP solution increases with ω_1, whereas the maximum NFV node utilization factor decreases with ω_1, because much imbalanced loading is penalized more than migrations.

Next, we evaluate the cost and time efficiency of the MIQCP and heuristic methods using a 256-node mesh network topology for the virtual resource pool. The comparison is done with fixed weights $\{\omega_1, \omega_2, \omega_3\} = \{0.4, 0.4, 0.2\}$ in (3.17), under the condition of $\omega_1\eta_U < \omega_2$. Eight groups of experiments are carried out with 10, 15, 20, 25, 30, 35, 40 and 45 SFCs randomly deployed in the virtual resource pool, respectively. At various traffic rates ranging from 200 to 740 packet/s, the total cost and running time for each group of experiments are evaluated. In each experiment, all the SFCs in the virtual resource pool have the same traffic rate, denoted by $\lambda(k)$. For example, in the experiment with 10 SFCs and at traffic rate 200 packet/s, all the

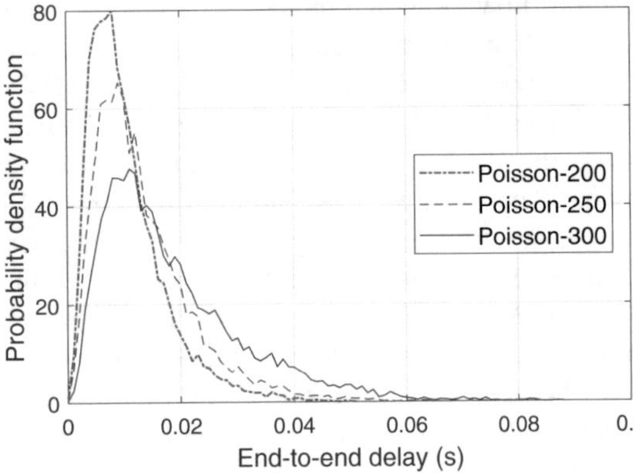

(a)

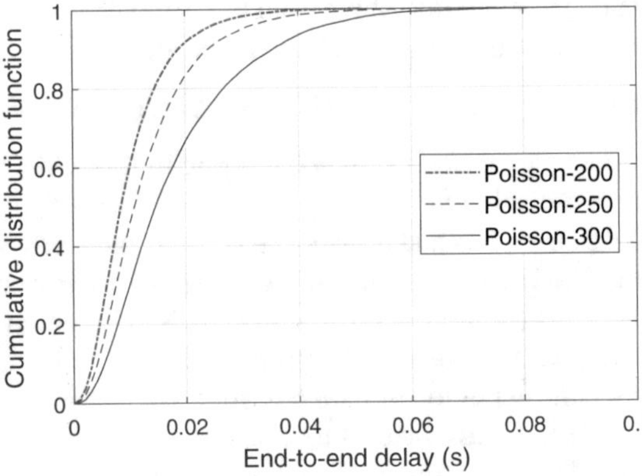

(b)

Fig. 3.23 E2E delay distribution for Poisson traffic models with different rates with flow migration. (**a**) Probability distribution function (PDF). (**b**) Cumulative function distribution (CDF)

Table 3.2 E2E delay parameters for Poisson traffic models with different rates with flow migration

Traffic models	Average E2E delay	P_{D_3}	P_{2*D_3}	P_{3*D_3}
Poisson-200	0.0110 s	91.73%	99.74%	100%
Poisson-250	0.0139 s	83.36%	98.56%	99.89%
Poisson-300	0.0189 s	66.67%	93.43%	99.22%

10 SFCs have a traffic rate of 200 packet/s. We set the initial step size, $\eta_{\Delta,0}$, and the precision, η^{ϵ}_{Δ}, in the heuristic algorithm as 0.1 and 0.0001 respectively.

The relationships between the average total cost and the number of SFCs ($|\mathcal{R}|$) for both the MIQCP and heuristic solutions are illustrated in Fig. 3.25, both showing an increasing trend with the increase of $|\mathcal{R}|$. The reason is that more SFCs increase the overall resource usage and create more overloaded NFV nodes, especially when the SFCs initially share common NFV nodes with each other at high traffic rates.

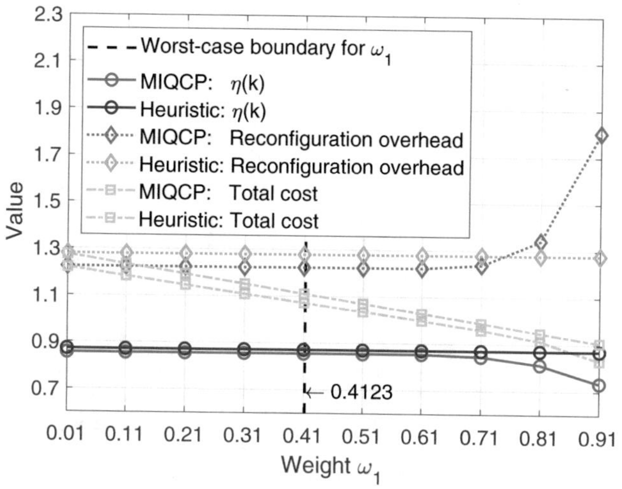

Fig. 3.24 Costs with respect to weight ω_1 in objective function, for $\omega_2 = 2\omega_3$

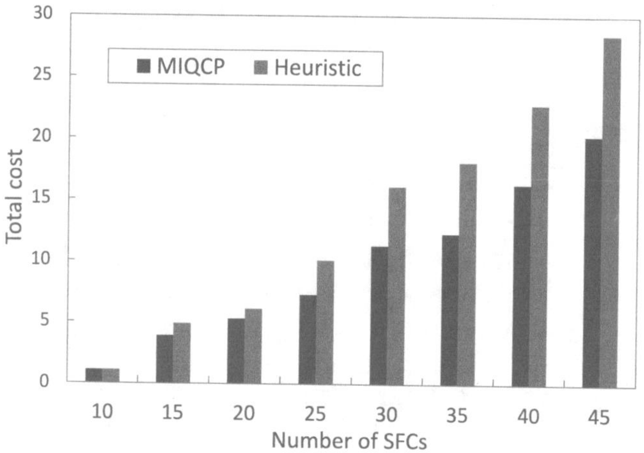

Fig. 3.25 Total cost with respect to the number of SFCs

To satisfy the QoS requirements of all SFCs and balance the resource utilization among the NFV nodes, more VNFs tend to migrate to alternative NFV nodes which incurs more migration cost. Moreover, due to increased overall traffic load with more SFCs, the maximum NFV node utilization ratio tends to be higher after flow migration.

The average running time with respect to $|\mathcal{R}|$ for both the MIQCP and heuristic solutions are illustrated in Fig. 3.26. Due to the NP-hardness of the MIQCP problem, the optimal MIQCP solution cannot be obtained in polynomial time, thus it cannot scale with the network scale. That's why we see an almost exponential increasing trend for the running time of the MIQCP solution as $|\mathcal{R}|$ increases. For the heuristic

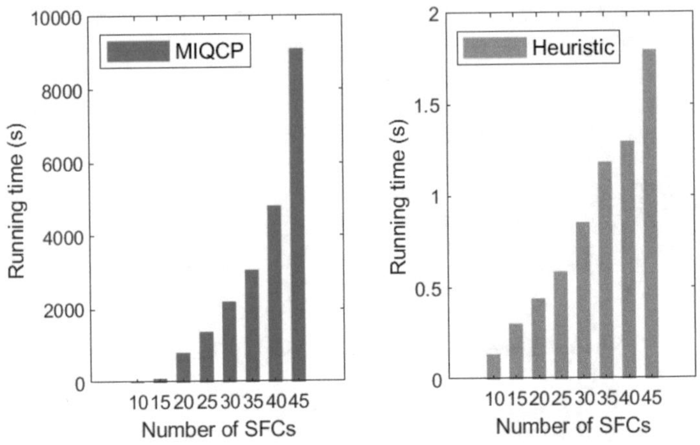

Fig. 3.26 Running time with respect to the number of SFCs

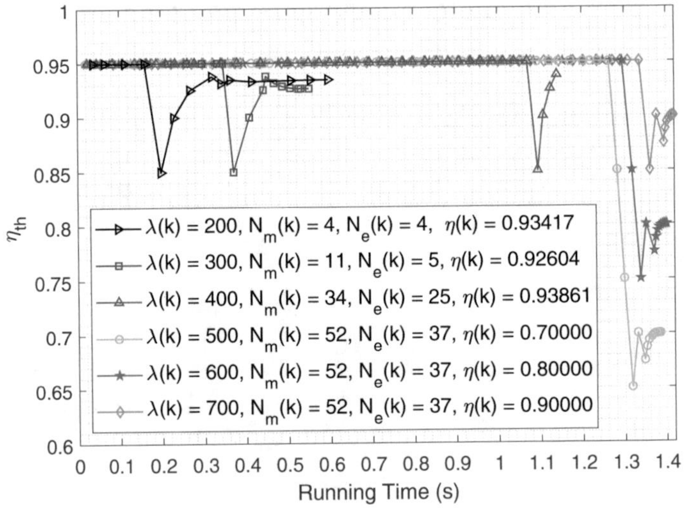

Fig. 3.27 Threshold update in the heuristic algorithm, for $\eta_U = 0.95$

solution, the increasing trend of the running time with the increase of $|\mathcal{R}|$ is much less significant, demonstrating the time efficiency of the heuristic solution.

Convergence of Heuristic Algorithm

We plot the updating process of the threshold η_{th} in the 45-SFC experiments with different traffic rates $\lambda(k)$, to evaluate convergence of the heuristic algorithm. The results are shown in Fig. 3.27,in which each threshold updating curve corresponds to one traffic rate.

On each threshold updating curve, we observe that η_{th} initially remains η_U during several sequential migration decisions at the beginning. When no more migration is required, η_{th} starts to decrease with the initial step size $\eta_{\Delta,0} = 0.1$ for load balancing, until it reaches a turning point at the lower bound. The turning point corresponds to the last value of η_{th} during the consecutive decreasing stage, at which there is at least one overloaded NFV node with utilization factor larger than η_{th}. For the values of η_{th} before the turning point, no NFV nodes have a utilization factor exceeding η_{th}. After the turning point, a binary search for the value of η_{th} is performed, and the step size is reduced by half for each iteration until it reaches the required precision $\eta_{\Delta}^{\epsilon} = 0.0001$. With the increase of traffic rate to 500 packet/s, more migrations happen to gradually decouple the SFCs from each other, and more new virtual links are observed.

For each traffic rate, $\eta(k)$, $N_m(k)$, and $N_e(k)$ denote the maximum NFV node utilization factor, the total number of VNF migrations, and the total number of required new virtual links, after convergence. When the traffic rate grows larger than 500 packet/s, all SFCs are completely decoupled, with no resource sharing on NFV nodes, thus $N_m(k)$ and $N_e(k)$ are stabilized but $\eta(k)$ increases. When the traffic rate is smaller than 500 packet/s, $\eta(k)$ is close to η_U, but less migrations and new virtual links are generated, demonstrating a trade-off between load balancing and migration cost.

3.6 Summary

In this chapter, we study a delay-aware flow migration problem for embedded services with average E2E delay requirements. A mixed integer optimization problem is formulated to address the trade-off between load balancing and migration cost, which non-convex and intractable in optimization solvers due to multiple quadratic constraints. By replacing each quadratic constraint by a group of tractable linear or quadratic cone constraints, the problem is reformulated to an MIQCP optimization problem. If feasible, the MIQCP problem has at least one optimum being the optimum of the original problem. Numerical results show that more traffic can be accommodated from services with average E2E delay guarantee, if flow migrations are allowed. Moreover, a flow migration strategy which balances between load balancing and migration cost reduction achieves medium load balancing, as compared with flow migration strategies with a focus on either goal. Nevertheless, it achieves approximately as good performance in terms of the migration cost as a flow migration strategy emphasizing on migration reduction. This result indicates the benefit of joint consideration of the two goals. A performance comparison between the optimal MIQCP and low-complexity heuristic solutions demonstrates the effectiveness and time efficiency of the heuristic solution [19–21].

References

1. Bhamare, D., Samaka, M., Erbad, A., Jain, R., Gupta, L., Chan, H.A.: Optimal virtual network function placement in multi-cloud service function chaining architecture. Comput. Commun. **102**, 1–16 (2017)
2. Ben Jemaa, F., Pujolle, G., Pariente, M.: Analytical models for QoS-driven VNF placement and provisioning in wireless carrier cloud. In: Proc. 19th ACM International Conf. on Modeling, Analysis and Simulation of Wireless and Mobile Systems, pp. 148–155 (2016)
3. Shin, M., Chong, S., Rhee, I.: Dual-resource TCP/AQM for processing-constrained networks. IEEE/ACM Trans. Netw. **16**(2), 435–449 (2008)
4. Rizzo, L., Carbone, M., Catalli, G.: Transparent acceleration of software packet forwarding using Netmap. In: Proc. IEEE INFOCOM, pp. 2471–2479 (2012)
5. Garzarella, S., Lettieri, G., Rizzo, L.: Virtual device passthrough for high speed VM networking. In: Proc. 11th ACM/IEEE Symp. Architectures for Networking and Communications Systems, pp. 99–110 (2015)
6. Dpdk: Data plane development kit. http://dpdk.org/. Accessed 14 July 2021
7. Ye, Q., Li, J., Qu, K., Zhuang, W., Shen, X., Li, X.: End-to-end quality of service in 5G networks—examining the effectiveness of a network slicing framework. IEEE Veh. Technol. Mag. **13**(2), 65–74 (2018)
8. Ye, Q., Zhuang, W., Li, X., Rao, J.: End-to-end delay modeling for embedded VNF chains in 5G core networks. IEEE Internet Things J. **6**(1), 692–704 (2019)
9. Emmerich, P., Raumer, D., Gallenmüller, S., Wohlfart, F., Carle, G.: Throughput and latency of virtual switching with Open vSwitch: a quantitative analysis. J. Netw. Syst. Manag. **26**(2), 314–338 (2018)
10. Ghaznavi, M., Khan, A., Shahriar, N., Alsubhi, K., Ahmed, R., Boutaba, R.: Elastic virtual network function placement. In: Proc. IEEE CloudNet, pp. 255–260 (2015)
11. Xia, J., Pang, D., Cai, Z., Xu, M., Hu, G.: Reasonably migrating virtual machine in NFV-featured networks. In: Proc. IEEE Conf. Computer and Information Technology, pp. 361–366 (2016)
12. Zhang, B., Zhang, P., Zhao, Y., Wang, Y., Luo, X., Jin, Y.: Co-scaler: cooperative scaling of software-defined NFV service function chain. In: Proc. IEEE Conf. Network Function Virtualization and Software Defined Networks, pp. 33–38 (2016)
13. Tang, H., Zhou, D., Chen, D.: Dynamic network function instance scaling based on traffic forecasting and VNF placement in operator data centers. IEEE Trans. Parallel Distrib. Syst. **30**(3), 530–543 (2019)
14. Luo, Z., Wu, C., Li, Z., Zhou, W.: Scaling geo-distributed network function chains: a prediction and learning framework. IEEE J. Sel. Areas Commun. **37**(8), 1838–1850 (2019)
15. Guo, L., Pang, J., Walid, A.: Dynamic service function chaining in SDN-enabled networks with middleboxes. In: Proc. IEEE ICNP, pp. 1–10 (2016)
16. Liu, J., Lu, W., Zhou, F., Lu, P., Zhu, Z.: On dynamic service function chain deployment and readjustment. IEEE Trans. Netw. Serv. Manag. **14**(3), 543–553 (2017)
17. Eramo, V., Miucci, E., Ammar, M., Lavacca, F.G.: An approach for service function chain routing and virtual function network instance migration in network function virtualization architectures. IEEE/ACM Trans. Netw. **25**(4), 2008–2025 (2017)
18. Garey, M.R., Johnson, D.S.: Computers and Intractability: A Guide to the Theory of NP-Completeness. WH Freeman, New York (1978)
19. Qu, K., Zhuang, W., Ye, Q., Shen, X., Li, X., Rao, J.: Dynamic flow migration for embedded services in SDN/NFV-enabled 5G core networks. IEEE Trans. Commun. **68**(4), 2394–2408 (2020)
20. Qu, K., Zhuang, W., Ye, Q., Shen, X., Li, X., Rao, J.: Traffic engineering for service-oriented 5g networks with SDN-NFV integration. IEEE Netw. **34**(4), 234–241 (2020)
21. Qu, K., Zhuang, W., Ye, Q., Shen, X., Li, X., Rao, J.: Delay-aware flow migration for embedded services in 5G core networks. In: Proc. IEEE ICC, pp. 1–6 (2019)

Chapter 4
Dynamic VNF Resource Scaling and Migration: A Machine Learning Approach

4.1 System Model

4.1.1 Network Model

We consider a local network segment and focus on one VNF in a service function chain (or VNF chain), with an incoming subflow from its upstream VNF, and an outgoing subflow towards its downstream VNF. The VNF can be placed at an NFV node in a candidate set, $\mathcal{N}_C$, with an initial placement at NFV node $n_0 \in \mathcal{N}_C$. For each NFV node $n \in \mathcal{N}_C$, it can be dynamically deployed with other VNFs from different services with a time-varying traffic load.

The traffic from VNFs other than the considered VNF is collectively referred to as background traffic. Define the background resource utilization factor of NFV node $n \in \mathcal{N}_C$ as the fraction of CPU computing resources occupied by the background traffic at NFV node $n \in \mathcal{N}_C$. With the dynamic VNF placement and traffic fluctuations of other VNFs, the background resource utilization factor of NFV node $n \in \mathcal{N}_C$ changes with time. The network scenario is illustrated in Fig. 4.1, where there are 5 candidate NFV nodes in set $\mathcal{N}_C$, and the considered VNF is currently placed at the upper left candidate NFV node.

Packet processing at the VNF has a stochastic delay requirement, requiring that the delay violation probability should not exceed an upper limit, i.e., $\Pr(\mathsf{d} > \mathsf{D}) \le \varepsilon$, where d is a random variable denoting the experienced VNF packet processing (including queueing) delay, D is the delay bound, and ε is the maximum delay violation probability. As the background traffic or the traffic of the considered VNF changes, the considered VNF can migrate to alternative NFV nodes in $n \in \mathcal{N}_C$, with resource scaling, to satisfy the stochastic delay requirement.

W. Zhuang, K. Qu, *Dynamic Resource Management in Service-Oriented Core Networks*, Wireless Networks, https://doi.org/10.1007/978-3-030-87136-9_4

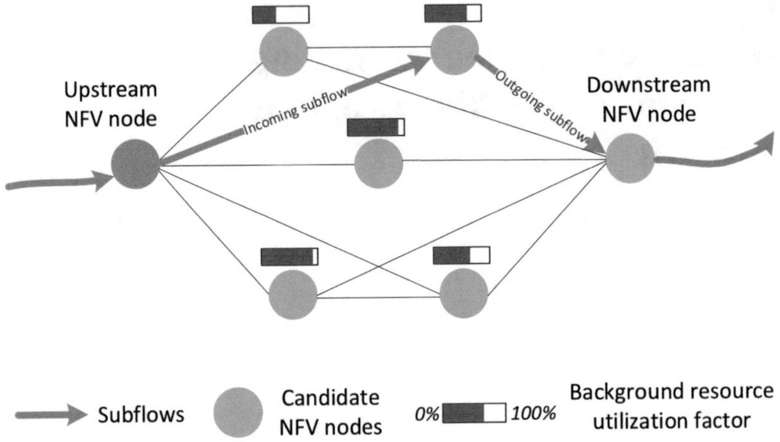

Fig. 4.1 An illustration of network scenario

4.1.2 *Traffic Model*

Consider a real-world traffic trace as the traffic input of the considered VNF. The packet arrival timestamps are available. With such packet-level information, we can derive discrete traffic time series in different timescales, to represent the multi-timescale traffic dynamics.

4.1.2.1 Multi-Timescale Time Series

Given a time interval length, time is partitioned into non-overlapping time intervals. For each time interval, the starting and ending times are known, so the total number of packet arrivals during the time interval, referred to as the traffic sample in the time interval, can be calculated based on the packet arrival timestamps in the traffic trace. All the traffic samples constitute a discrete traffic time series for the given time interval length. Figure 4.2 illustrates the relationship between traffic time series and packet arrival timestamps, for a given time interval length.

We consider two time series for the same traffic trace, one in a medium timescale with medium interval length (in second) equal T_M (e.g., 20 s), and one in a small timescale with small interval length (in second) equal T_S (e.g., 0.1 s). Let $\boldsymbol{x}_M$ denote the time series in medium timescale, given by

$$\boldsymbol{x}_M = [x_M(0), x_M(1), \cdots, x_M(i), \cdots] \tag{4.1}$$

where i (≥ 0) is an index for the medium time interval and $x_M(i)$ is the i-th traffic sample in medium timescale, representing the number of packet arrivals in the i-th medium time interval. Let $\boldsymbol{x}_M[i : i']$ denote a series of traffic samples between

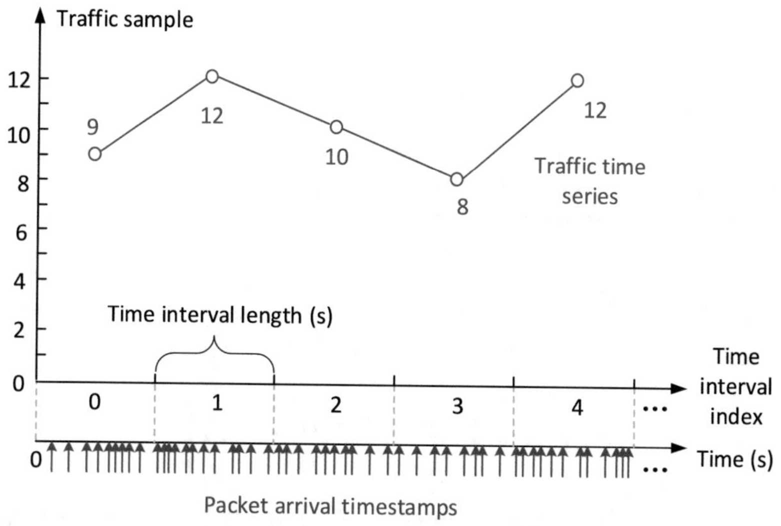

Fig. 4.2 An illustration of the relationship between traffic time series and packet arrival timestamps, for given time interval length

medium time intervals i and i' (inclusive), given by

$$\boldsymbol{x}_M[i:i'] = \left[x_M(i), x_M(i+1), \cdots, x_M(i'-1), x_M(i')\right], \quad i' > i. \tag{4.2}$$

Similarly, let $\boldsymbol{x}_S$ denote the traffic time series in small timescale, represented as

$$\boldsymbol{x}_S = [x_S(0), x_S(1), \cdots, x_S(t), \cdots] \tag{4.3}$$

where t (≥ 0) is an index for the small time interval and $x_S(t)$ is the t-th traffic sample in small timescale, representing the number of packet arrivals in the t-th small time interval. Let $\boldsymbol{x}_S[t:t']$ denote a series of traffic samples between small time intervals t and t' (inclusive), given by

$$\boldsymbol{x}_S[t:t'] = \left[x_S(t), x_S(t+1), \cdots, x_S(t'-1), x_S(t')\right], \quad t' > t. \tag{4.4}$$

For the small timescale, denote $\mathsf{A}(t)$ as the cumulative number of packet arrivals before small time interval t, which is calculated as

$$\mathsf{A}(t) = \begin{cases} \sum_{t'=1}^{t} x_S(t'-1), & t \geq 1 \\ 0, & t = 0. \end{cases} \tag{4.5}$$

Let Λ denote the long-term average traffic rate of the traffic trace in packet/s. Then, the following relationship between the medium-timescale and small-timescale traffic time series holds, given by

$$\Lambda = \lim_{i'\to\infty} \frac{1}{i'} \sum_{i=0}^{i'} \frac{x_M(i)}{T_M} = \lim_{t'\to\infty} \frac{1}{t'} \sum_{t=0}^{t'} \frac{x_S(t)}{T_S} \tag{4.6}$$

where $\frac{x_M(i)}{T_M}$ and $\frac{x_S(t)}{T_S}$ are the average traffic rates (in packet/s) in the i-th medium time interval and the t-th small time interval, respectively.

Assume that the medium time interval length, T_M, is multiples of the small time interval length, T_S. Then, each medium-timescale traffic sample corresponds to $\frac{T_M}{T_S}$ small-timescale traffic samples. Specifically, the traffic sample in the i-th medium time interval, i.e., $x_M(i)$, corresponds to a small-timescale traffic time series between small time intervals $\frac{iT_M}{T_S}$ and $\left(\frac{(i+1)T_M}{T_S} - 1\right)$, represented as

$$x_M(i) \Rightarrow \boldsymbol{x}_S\left[\frac{iT_M}{T_S} : \left(\frac{(i+1)T_M}{T_S} - 1\right)\right]. \tag{4.7}$$

In general, a medium-timescale traffic time series can be mapped to a small-timescale traffic time series within the same time duration, represented as

$$\boldsymbol{x}_M[i : i'] \Rightarrow \boldsymbol{x}_S\left[\frac{iT_M}{T_S} : \left(\frac{(i'+1)T_M}{T_S} - 1\right)\right]. \tag{4.8}$$

4.1.2.2 Stationary Traffic Segments with Unknown Change Points

As illustrated in Fig. 1.7, the real-world network traffic exhibits a daily periodic pattern in large timescale, showing peaks in the days and valleys in the nights, which demonstrates the traffic non-stationarity [1]. Hence, both the traffic time series in medium and small timescales, $\boldsymbol{x}_M$ and $\boldsymbol{x}_S$, are non-stationary time series.

Assume that the non-stationary traffic trace can be partitioned into consecutive stationary traffic segments with unknown change points in time. For each stationary traffic segment between two neighboring change points, traffic statistics such as mean and variance do not change. Correspondingly, the non-stationary traffic time series in either medium or small timescale can be partitioned into non-overlapping stationary traffic time series segments with corresponding change points in the indexes of traffic samples.

Let integer k (≥ 0) indicate the k-th stationary traffic segment in the non-stationary traffic trace. The starting time of the k-th stationary traffic segment is referred to as the k-th change point in time. Consider that there exists a change point detection algorithm which detects the change points based on traffic statistical changes in the medium-timescale time series, whose outputs are the changes points in the indexes of medium-timescale traffic samples. The index of the first traffic sample in the k-th stationary traffic time series in medium timescale is referred to as the k-th change point in medium timescale, denoted by $C_M(k)$. Then, $x_M(C_M(k))$ is

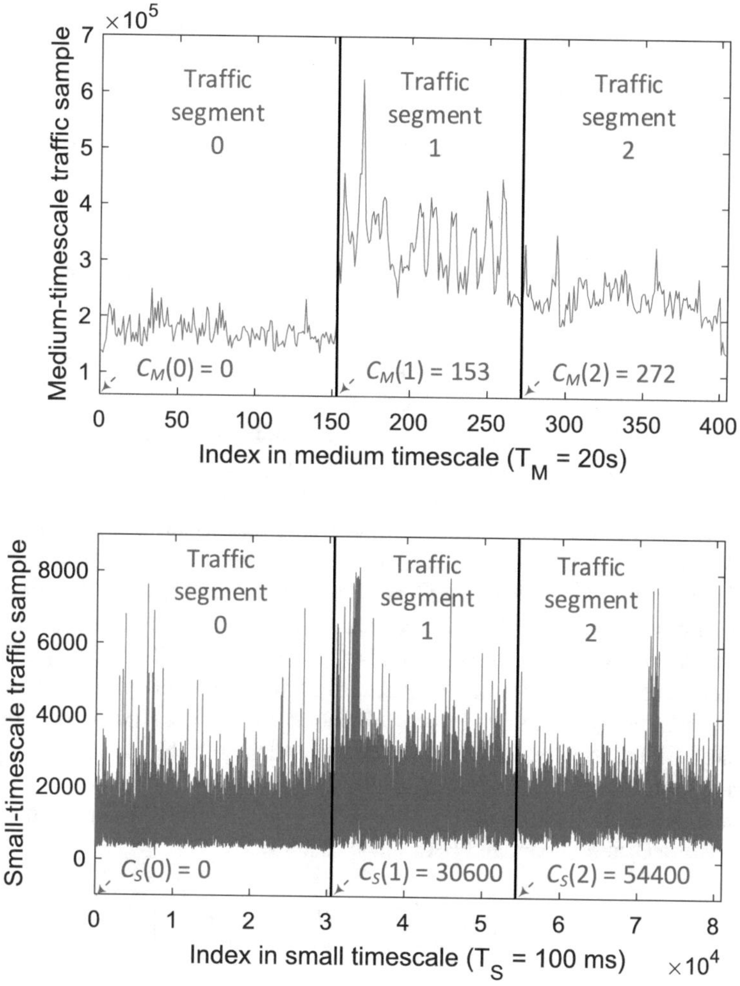

Fig. 4.3 An illustration of non-stationary traffic in different timescales

the first medium-timescale traffic sample of the k-th stationary traffic segment. We have $C_M(0) = 0$ to indicate the beginning of the time line.

Correspondingly, the k-th change point in small timescale, $C_S(k)$, denoting the index of the first small-timescale traffic sample of the k-th stationary traffic segment, is given by

$$C_S(k) = \frac{C_M(k)T_M}{T_S}. \tag{4.9}$$

Figure 4.3 shows the medium and small timescale time series for a segment of the non-stationary traffic trace, consisting of three stationary traffic segments.

The medium and small time interval lengths are $T_M = 20$ s and $T_S = 100$ ms respectively. The change points between stationary traffic segments are indicated in the figure. For example, the 2-nd change point in medium timescale, $C_M(2)$, is the index of the first medium-timescale traffic sample in traffic segment 2, which is $C_M(2) = 272$ in the example. Correspondingly, the 2-nd change point in small timescale is given by $C_S(2) = 200C_M(2) = 54400$.

4.1.2.3 Factional Brownian Motion for Each Stationary Traffic Segment

For each stationary traffic segment in the non-stationary traffic trace, the traffic statistics are unchanged. We need a traffic model to characterize the traffic statistics.

As we know, traffic arrivals of a VNF in the core network are from a service-level aggregated flow, which is an aggregation of individual traffic flows of different users subscribed to the service. The aggregation level of the core network traffic is high, which makes Gaussian traffic approximation work well beyond a timescale of around 100 ms [2]. Gaussianity of a certain distribution can be checked by quantile-quantile (Q-Q) plot versus a standard Gaussian distribution [3]. Beyond Gaussianity, the real-world network traffic in core network usually has properties of self-similarity and long-range dependence (LRD) [2, 4]. Fractional Brownian motion (fBm) is a Gaussian process with properties such as self-similarity and long-range dependence (LRD) for certain parameters, which comply with the properties of real-world network traffic. Hence, we adopt the fBm traffic model for each stationary traffic segment, based on which the statistical traffic parameters of each stationary traffic segment can be characterized.

A standard fBm process $\{Z_s(t), t = 0, 1, \cdots\}$ is a centered Gaussian process with $Z_s(0) = 0$, zero mean, and covariance function

$$\psi_{Z_s}(t_1, t_2) = \frac{1}{2}\left(t_1^{2H} + t_2^{2H} - |t_1 - t_2|^{2H}\right) \tag{4.10}$$

where $H \in (0, 1)$ is Hurst parameter [4]. For $H \in [0.5, 1)$, the fBm process is both self-similar and LRD.

For a stationary traffic segment starting from time 0, the cumulative number of packet arrivals before the t-th time unit can be modeled as a general fBm process, denoted by $\{Z(t), t = 0, 1, \cdots\}$, which is represented as

$$Z(t) = \lambda t + \sigma Z_s(t) \tag{4.11}$$

where $\lambda = \mathbb{E}(\frac{Z(t)}{t})$ is the mean of packet arrivals in a time unit, and σ is the standard deviation of packet arrivals in a time unit [4]. We consider a general fBm process in the small timescale. Hence, a time unit here corresponds to a small time interval. The covariance function of the general fBM process, $Z(t)$, is scaled by σ^2 and given by

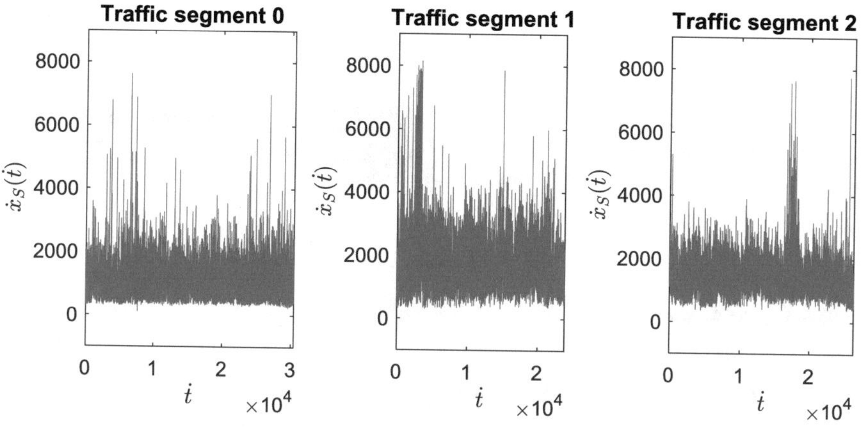

Fig. 4.4 Traffic segments with shifted time lines in small timescale

$$\psi_Z(t_1, t_2) = \frac{\sigma^2}{2}\left(t_1^{2H} + t_2^{2H} - |t_1 - t_2|^{2H}\right). \tag{4.12}$$

Hence, a general fBm process is characterized by three traffic parameters, i.e., $\{\lambda, \sigma, H\}$.

As illustrated in Fig. 4.3, only traffic segment 0 of a non-stationary traffic trace starts from the beginning of the time line. To model each traffic segment as a general fBm process, we consider different shifted time lines for each stationary traffic segment. With the time line shift, each stationary traffic segment starts from time 0 in its own shifted time line.

Specifically, for the k-th stationary traffic segment, we consider a shifted discrete time line, $\dot{t}$, in small timescale, starting at the beginning of the k-th stationary traffic segment and ending at the end of the k-th stationary traffic segment. Hence, the shifted discrete time line for the k-th stationary traffic segment is $\dot{t}_k = t - C_S(k), C_S(k) \leq t \leq C_S(k+1) - 1$. Accordingly, the range of $\dot{t}_k$ is $0 \leq \dot{t} \leq C_S(k+1) - C_S(k) - 1$. For clarity, we neglect the subscript k in $\dot{t}_k$, and represent it as $\dot{t}$. For the k-th stationary traffic segment, the number of packet arrivals in the $\dot{t}$-th shifted small time interval is represented as $\dot{x}_S(\dot{t}) = x_S(t - C_S(k)), C_S(k) \leq t \leq C_S(k+1) - 1$. Fig. 4.4 shows the three traffic segments with their shifted time lines in small timescale.

The cumulative number of packet arrivals in the k-th stationary traffic segment before $\dot{t}$, denoted by $\dot{\mathsf{A}}_k(\dot{t})$, is calculated as

$$\dot{\mathsf{A}}_k(\dot{t}) = \begin{cases} \sum_{\dot{t}'=1}^{\dot{t}} \dot{x}_S(\dot{t}' - 1), & 1 \leq \dot{t} \leq C_S(k+1) - C_S(k) - 1 \\ 0, & \dot{t} = 0. \end{cases} \tag{4.13}$$

We model $\dot{\mathsf{A}}_k(\dot{t})$ as a general fBm process with traffic parameters $\{\lambda\,(k), \sigma\,(k), H\,(k)\}$.

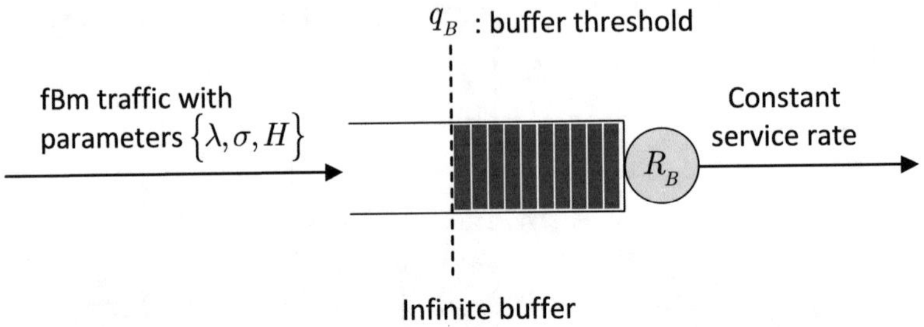

Fig. 4.5 Queueing model at the VNF with fBm traffic input

4.1.3 Resource Provisioning Model

Consider an fBm traffic input with parameters $\{\lambda, \sigma, H\}$ to an infinite buffer, with a constant service rate of R_S packets per small time interval, as illustrated in Fig. 4.5. The buffer overflow probability, i.e., the probability that queue length q is beyond a threshold q_B, is approximately given by Cheng et al. [4], Kim and Shroff [5]

$$\Pr(q > q_B) \simeq \exp\left(-\inf_{t\geq 0} \frac{[q_B + (\mathsf{R}_S - \lambda)t]^2}{2\sigma^2 t^{2H}}\right) \tag{4.14}$$

which has been shown accurate even for a small value of q_B by simulation studies. Correspondingly, the delay violation probability can be approximated by

$$\Pr(\mathsf{d}_S > \mathsf{D}_S) \simeq \exp\left(-\inf_{t\geq 0} \frac{[\mathsf{R}_S\mathsf{D}_S + (\mathsf{R}_S - \lambda)t]^2}{2\sigma^2 t^{2H}}\right) \tag{4.15}$$

where $\mathsf{d}_S = \frac{\mathsf{d}}{T_S}$ is the random VNF packet processing delay in number of small time intervals, and $\mathsf{D}_S = \frac{\mathsf{D}}{T_S}$ is the delay bound at the VNF in number of small time intervals.

The stochastic delay requirement at the VNF is described by a delay violation probability upper limit, i.e., $\Pr(\mathsf{d} > \mathsf{D}) \leq \varepsilon$, which is equivalent to $\Pr(\mathsf{d}_S > \mathsf{D}_S) \leq \varepsilon$. To provide such a probabilistic QoS guarantee to the VNF with service rate in packet per small time interval, denoted by $\mathsf{R}_{S,min}$, we should find

$$\min\left\{\mathsf{R}_S \mid \forall t \geq 0, \left[\mathsf{R}_S\mathsf{D}_S + (\mathsf{R}_S - \lambda)t\right]^2 \geq (-2\log\varepsilon)\sigma^2 t^{2H}\right\} \tag{4.16}$$

which leads to

$$\mathsf{R}_{S,min} = \sup_{t\geq 0} \frac{\lambda t + \sqrt{-2\log\varepsilon}\,\sigma t^H}{t + \mathsf{D}_S}. \tag{4.17}$$

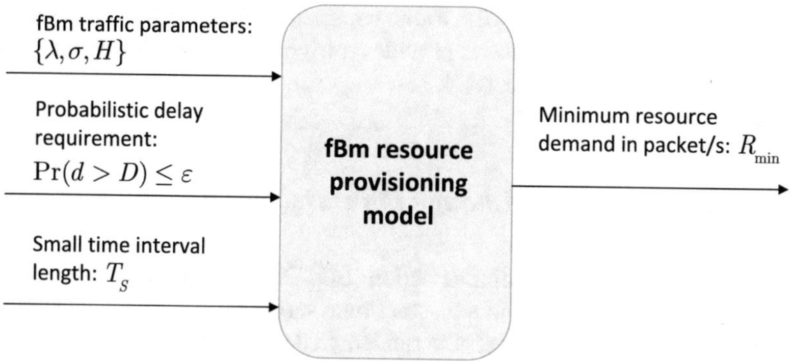

Fig. 4.6 A diagram for the fBm resource provisioning model

The optimal value of t achieving the supremum in (4.17), i.e., $t^{\star}$, can be obtained by setting the derivative of $\frac{\lambda t+\sqrt{-2\log\varepsilon}\sigma t^{H}}{t+\mathsf{D}_S}$ with respect to t to zero, i.e.,

$$\frac{\sqrt{-2\log\varepsilon}\sigma\mathsf{D}_S H {t^{\star}}^{H-1}+\sqrt{-2\log\varepsilon}\sigma(H-1){t^{\star}}^{H}+\lambda\mathsf{D}_S}{(t^{\star}+\mathsf{D}_S)^2}=0. \tag{4.18}$$

The numerical result of $t^{\star}$ can be obtained. With the optimal value $t^{\star}$, we can calculate the minimum service rate at a VNF to support an fBm traffic input with parameters $\{\lambda, \sigma, H\}$ with probabilistic delay guarantee, given by

$$\mathsf{R}_{S,min}=\frac{\lambda t^{\star}+\sqrt{-2\log\varepsilon}\sigma {t^{\star}}^{H}}{t+\mathsf{D}_S}. \tag{4.19}$$

Then, the corresponding minimum resource demand of a VNF in packet/s, denoted by R_{min}, is given by

$$\mathsf{R}_{min}=\frac{\mathsf{R}_{S,min}}{T_S}. \tag{4.20}$$

Equations (4.17)–(4.20) are collectively referred to as the fBm resource provisioning model. The inputs and outputs of the model is summarized in Fig. 4.6.

4.2 Resource Demand Prediction for Non-stationary Traffic

Since traffic statistics change across different stationary traffic segments, the amount of computing resources allocated to the VNF for probabilistic QoS guarantee, i.e., $\Pr(\mathsf{d} > \mathsf{D}) \leq \varepsilon$, should be dynamically adjusted. Here, a change-point-driven traffic parameter learning and resource demand prediction scheme is used, to predict resource demands from learned fBm traffic parameters of stationary traffic segments between detected change points. It provides a triggering signal for dynamic VNF migration to be discussed in Sect. 4.4.

4.2.1 Bayesian Online Change Point Detection

The Bayesian online change point detection (BOCPD) algorithm is first introduced in [6], to detect change points in a discrete time series using a statistical approach. Central to the BOCPD algorithm is the run length denoted by l. A run is defined as a traffic segment with the same statistics. The run length at each time step represents the number of samples before the current sample within the same run. Online inference about the run length is performed at every time step, given a conditional prior distribution over the run length and an underlying predictive model. The posterior probability distribution of run length at each time step is inferred, based on which change points can be detected.

We use the medium-timescale traffic time series for change point detection. Assume that the traffic samples in the non-stationary medium-timescale time series, $\boldsymbol{x}_M$, are from i.i.d Gaussian distribution $\mathcal{N}(\mu, \nu^2)$ with unknown (and perhaps changing) mean μ and standard deviation ν.[1] We use the BOCPD algorithm to detect statistical changes in mean and variance of the non-stationary time series in medium timescale. Hence, a medium time interval corresponds to a time step in the BOCPD algorithm.

To explain how the BPCPD algorithm works, we first formally define run length, then derive the posterior distribution of run length at each time step given the current traffic sample observations, and finally define two thresholds with which deterministic change points can be detected.

4.2.1.1 Run Length

The run length at the i-th time step, denoted by l_i, is a random variable. It represents the number of traffic samples before the i-th traffic sample, $x_M(i)$, within the same run. The random variable l_i takes values from $\{0, 1, \cdots, i\}$, as illustrated in Fig. 4.7. If $l_i = 0$, the i-th traffic sample, $x_M(i)$, has no traffic samples before it within the same run. Hence, the i-th traffic sample must be the first traffic sample in a new traffic segment. If $l_i = i$, all the traffic samples from the very beginning of the time line to the current traffic sample, $x_M(i)$, belong to the same traffic segment. The l_i traffic samples before traffic sample $x_M(i)$ in the same run must be the most recent

[1] Note that the i.i.d Gaussian assumption is used to detect change points. For traffic parameter learning, we do not rely on such an assumption.

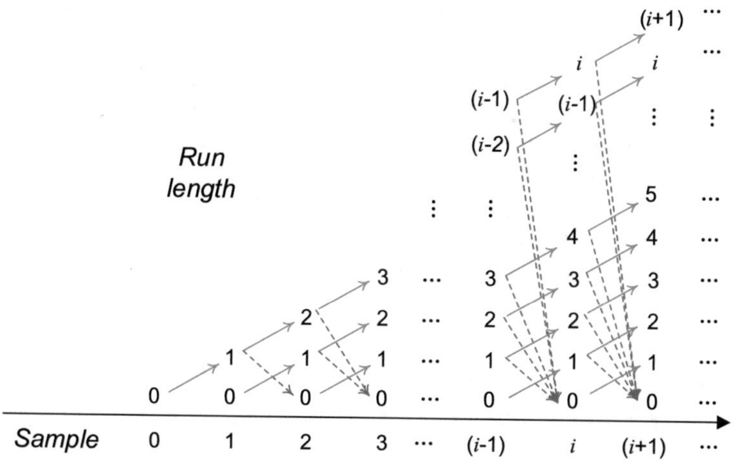

Fig. 4.7 An illustration of run length growth

l_i traffic samples, i.e., $x_M[(i-l_i):i-1]$. For example, if $l_i=2$, $x_M(i-2)$ and $x_M(i-1)$ are in the same run with $x_M(i)$.

For notation simplification, we omit the subscript M denoting the medium timescale. We use x_i and $\boldsymbol{x}_i$ to denote $x_M(i)$ and $\boldsymbol{x}_M[0:i]$, respectively. We also use $\boldsymbol{x}_i^{(l)}$ to denote $x_M[(i-l_i):i]$, which is a time series in the same run before the $(i+1)$-th traffic sample, given the run length l_i at time step i.

From time step $(i-1)$ to i, the run length either increases by 1 or resets to 0, as illustrated in Fig. 4.7. Hence, the conditional probability of $\Pr(l_i|l_{i-1})$ is nonzero only at $l_i=l_{i-1}+1$ and $l_i=0$. Assume that the change point generation at each time step follows a prior Bernoulli distribution with probability $\frac{1}{\bar{l}}$. Then, the number of time steps between two neighboring change points follows a prior geometric distribution with parameter $\frac{1}{\bar{l}}$, and the average run length between two neighboring change points is $\bar{l}$. If time step i is a change point, the run length l_i is reset to 0; otherwise, l_i is increased from l_{i-1} by 1. Hence, the conditional prior probability distribution for the run length at time step i, denoted by $\Pr(l_i|l_{i-1})$, is given by

$$\Pr(l_i|l_{i-1}) = \begin{cases} 1-(1/\bar{l}), & \text{if } l_i = l_{i-1}+1 \\ 1/\bar{l}, & \text{if } l_i = 0 \\ 0, & \text{otherwise.} \end{cases} \tag{4.21}$$

The conditional prior has nonzero mass at only two outcomes, i.e., the run length either grows by 1, or resets to 0.

4.2.1.2 Posterior Run Length Distribution

To derive the posterior distribution of run length given the current traffic sample observations at time step i, i.e., $\Pr(l_i = i'|\boldsymbol{x}_i)$, let's first characterize the relationship between the joint probability distribution of run length and observed time series for two consecutive time steps, e.g., time steps $i-1$ and i.

The joint probability of run length and observed time series at time step i, i.e., $\Pr(l_i, \boldsymbol{x}_i)$, is updated recursively from the joint probability at the previous time step, i.e., $\Pr(l_{i-1}, \boldsymbol{x}_{i-1})$, for $i \geq 1$, given by

$$\overbrace{\Pr(l_i, \boldsymbol{x}_i)}^{i\text{-th iteration}} = \sum_{l_{i-1}} \underbrace{\Pr(l_i|l_{i-1})}_{\text{conditional prior on run length}} \underbrace{\Pr(x_i|l_{i-1}, \boldsymbol{x}_{i-1}^{(l)})}_{\text{predictive model}} \overbrace{\Pr(l_{i-1}, \boldsymbol{x}_{i-1})}^{(i-1)\text{-th iteration}}, \quad \forall i \geq 1. \tag{4.22}$$

The underlying condition for (4.22) is that run length l_i is independent of $\boldsymbol{x}_i$, given l_{i-1}. The condition is true under the assumption of the conditional prior on run length in (4.21). According to (4.22), the joint probability distribution for each time step is updated recursively from an initialization at time step 0, which is $\Pr(l_0 = 0, x_0) = 1$, for any observed value of x_0. With such an initialization, the joint probability represents a relative likelihood. At each time step, the joint probability distribution is updated based on two models: conditional prior distribution on run length, and predictive model for traffic sample.

- The conditional prior on run length, i.e., $\Pr(l_i|l_{i-1})$, is a probability mass distribution with two outcomes, i.e., $l_i = l_{i-1} + 1$ and $l_i = 0$, as given in (4.21). Accordingly, there are two branches for $\Pr(l_i, \boldsymbol{x}_i)$, given by

$$\overbrace{\Pr(l_i, \boldsymbol{x}_i)}^{i\text{-th iteration}} = \begin{cases} \left(1 - \frac{1}{l}\right) \Pr\left(x_i|l_{i-1}, \boldsymbol{x}_{i-1}^{(l)}\right) \Pr\left(l_{i-1}, \boldsymbol{x}_{i-1}\right), & \text{if } l_i = l_{i-1} + 1 \\ \frac{1}{l} \sum_{l_{i-1}=0}^{i-1} \Pr\left(x_i|l_{i-1}, \boldsymbol{x}_{i-1}^{(l)}\right) \Pr\left(l_{i-1}, \boldsymbol{x}_{i-1}\right), & \text{if } l_i = 0. \end{cases} \tag{4.23}$$

- The predictive model, i.e., $\Pr(x_i|l_{i-1}, \boldsymbol{x}_{i-1}^{(l)})$, evaluates the probability of the future traffic sample at time step i equal x_i, given l_{i-1} and $\boldsymbol{x}_{i-1}^{(l)}$ (i.e., $x_M[(i - 1 - l_{i-1}) : (i-1)]$. With a Gaussian-Inverse-Gamma prior on the unknown mean, μ, and variance, ν^2, of the i.i.d Gaussian distribution, the predictive model is described by a student-t distribution with mean $\mu_{i-1}^{(l)}$ and standard deviation $\nu_{i-1}^{(l)}$, based on conjugate Bayesian analysis. For each possible value of l_{i-1}, both $\mu_{i-1}^{(l)}$ and $\nu_{i-1}^{(l)}$ take different values.

Conjugate Bayesian Analysis

At time step $i-1$, time series $\boldsymbol{x}_{i-1}^{(l)} = x_M[(i-1-l_{i-1}):(i-1)]$ belong to the same run, given run length l_{i-1}. Under the i.i.d Gaussian assumption, time series $\boldsymbol{x}_{i-1}^{(l)}$ are composited by traffic samples from Gaussian distribution $\mathcal{N}(\mu, \nu^2)$ with unknown mean μ and standard deviation ν.

A Normal-Inverse-Gamma prior is placed on μ and ν^2, of the i.i.d Gaussian distribution, given by

$$\Pr\left(\mu, \nu^2\right) \sim \mathcal{N}\left(\mu|\mu_0, \frac{\nu^2}{\kappa_0}\right) \mathcal{IG}\left(\nu^2|\alpha_0, \beta_0\right) \tag{4.24}$$

where $\{\mu_0, \kappa_0, \alpha_0, \beta_0\}$ are prior parameters of the Normal-Inverse-Gamma distribution. With the online arrivals of traffic sample observations, Conjugate Bayesian analysis [7, 8] gives a Normal-Inverse-Gamma posterior on μ and ν^2 given the observed time series at each time step.

Suppose the posterior on μ and ν^2 given l_{i-2} and $\boldsymbol{x}_{i-2}^{(l)}$ at time step $i-2$ $(i \geq 2)$ is represented as

$$\Pr\left(\mu, \nu^2|l_{i-2}, \boldsymbol{x}_{i-2}^{(l)}\right) \sim \mathcal{N}\left(\mu\middle|\mu_{i-2}^{(l)}, \frac{\nu^2}{\kappa_{i-2}^{(l)}}\right) \mathcal{IG}\left(\nu^2|\alpha_{i-2}^{(l)}, \beta_{i-2}^{(l)}\right) \tag{4.25}$$

with parameters $\{\mu_{i-2}^{(l)}, \kappa_{i-2}^{(l)}, \alpha_{i-2}^{(l)}, \beta_{i-2}^{(l)}\}$, which are referred to as sufficient statistics corresponding to l_{i-2} and $\boldsymbol{x}_{i-2}^{(l)}$. At time step $i-2$, each possible value of run length l_{i-2} corresponds to a specific time series $\boldsymbol{x}_{i-2}^{(l)}$ with length $l_{i-2}+1$, thus corresponding to a different group of sufficient statistics.

Next, we discuss how the Normal-Inverse-Gamma posterior on μ and ν^2 is updated from time step $i-2$ to time step $i-1$, after observation x_{i-1} is available. There are two cases for the update, depending on the value of l_{i-1}.

- **Case 1:** If $l_{i-1} = l_{i-2}+1$, traffic sample x_{i-1} belongs to the same run as $\boldsymbol{x}_{i-2}^{(l)}$. In this case, one more observation of the Gaussian distribution $\mathcal{N}(\mu, \nu^2)$ is available. Based on conjugate Bayesian analysis, the sufficient statistics corresponding to l_{i-1} and $\boldsymbol{x}_{i-1}^{(l)}$ for $l_{i-1} = l_{i-2}+1$ are updated as

$$\mu_{i-1}^{(l)} = \frac{\kappa_{i-2}^{(l)}\mu_{i-2}^{(l)} + x_{i-1}}{\kappa_{i-2}^{(l)}+1} \tag{4.26a}$$

$$\kappa_{i-1}^{(l)} = \kappa_{i-2}^{(l)} + 1 \tag{4.26b}$$

$$\alpha_{i-1}^{(l)} = \alpha_{i-2}^{(l)} + \frac{1}{2} \tag{4.26c}$$

$$\beta_{i-1}^{(l)} = \beta_{i-2}^{(l)} + \frac{\kappa_{i-2}^{(l)} \left(x_{i-1} - \mu_{i-2}^{(l)} \right)^2}{2 \left(\kappa_{i-2}^{(l)} + 1 \right)}. \tag{4.26d}$$

With the updated sufficient statistics $\{\mu_{i-1}^{(l)}, \kappa_{i-1}^{(l)}, \alpha_{i-1}^{(l)}, \beta_{i-1}^{(l)}\}$, the posterior on μ and ν^2 given l_{i-1} and $\boldsymbol{x}_{i-1}^{(l)}$ is updated as

$$\Pr\left(\mu, \nu^2 | l_{i-1}, \boldsymbol{x}_{i-1}^{(l)}\right) \sim \mathcal{N}\left(\mu \middle| \mu_{i-1}^{(l)}, \frac{\nu^2}{\kappa_{i-1}^{(l)}}\right) \mathcal{IG}\left(\nu^2 | \alpha_{i-1}^{(l)}, \beta_{i-1}^{(l)}\right). \tag{4.27}$$

Based on the posterior on μ and ν^2, conjugate Bayesian analysis also gives a posterior predictive distribution for x_i given l_{i-1} and $\boldsymbol{x}_{i-1}^{(l)}$, i.e., $\Pr(x_i | l_{i-1}, \boldsymbol{x}_{i-1}^{(l)})$, which is described by a student-t distribution, represented as

$$\Pr\left(x_i | l_{i-1}, \boldsymbol{x}_{i-1}^{(l)}\right) \sim t_{2\alpha_{i-1}^{(l)}} \left(x_i \middle| \mu_{i-1}^{(l)}, \frac{\beta_{i-1}^{(l)} \left(\kappa_{i-1}^{(l)} + 1 \right)}{\kappa_{i-1}^{(l)} \alpha_{i-1}^{(l)}} \right) \tag{4.28}$$

where $\mu_{i-1}^{(l)}$ is the mean, $2\alpha_{i-1}^{(l)}$ is the degrees of freedom, and $\frac{\beta_{i-1}^{(l)}(\kappa_{i-1}^{(l)}+1)}{\kappa_{i-1}^{(l)}\alpha_{i-1}^{(l)}}$ is the scale. The standard deviation of the student-t distribution, denoted by $\nu_{i-1}^{(l)}$, is given by

$$\nu_{i-1}^{(l)} = \sqrt{\frac{\beta_{i-1}^{(l)} \left(\kappa_{i-1}^{(l)} + 1 \right)}{\kappa_{i-1}^{(l)} \left(\alpha_{i-1}^{(l)} - 1 \right)}}. \tag{4.29}$$

The student-t distribution has a heavier tail than the Gaussian distribution with the same mean and variance, thus it is more robust to outliers [8].

- **Case 2:** If $l_{i-1} = 0$, traffic sample x_{i-1} is the very beginning traffic sample in a new run. Then, we have $\boldsymbol{x}_{i-1}^{(l)} = x_{i-1}$ with length 1, and the corresponding sufficient statistics updated from the prior parameters of the Normal-Inverse-Gamma distribution, represented as

$$\mu_{i-1}^{(l)} = \frac{\kappa_0 \mu_0 + x_{i-1}}{\kappa_0 + 1} \tag{4.30a}$$

$$\kappa_{i-1}^{(l)} = \kappa_0 + 1 \tag{4.30b}$$

$$\alpha_{i-1}^{(l)} = \alpha_0 + \frac{1}{2} \tag{4.30c}$$

$$\beta_{i-1}^{(l)} = \beta_0 + \frac{\kappa_0 \left(x_{i-1} - \mu_0\right)^2}{2\left(\kappa_0 + 1\right)}. \tag{4.30d}$$

With the updated sufficient statistics $\{\mu_{i-1}^{(l)}, \kappa_{i-1}^{(l)}, \alpha_{i-1}^{(l)}, \beta_{i-1}^{(l)}\}$, the posterior on μ and ν^2 given $l_{i-1} = 0$ and $\boldsymbol{x}_{i-1}^{(l)} = x_{i-1}$ is given by (4.27), and the posterior predictive model for time step i is given by (4.28).

For $i \geq 2$, we have discussed (1) how the posterior on μ and ν^2 is updated from time step $i-2$ to time step $i-1$ with new observation of x_{i-1} and for different values of l_{i-1}, and (2) how the traffic sample at time step i is predicted, based on the posterior on μ and ν^2 at time step $i-2$. The missing part is how the posterior on μ and ν^2 at time step 0 is obtained. Since traffic sample x_0 is at the beginning of the whole non-stationary time series, run length l_0 is equal to 0, as illustrated in Fig. 4.7. Hence, the sufficient statistics at time step 0 is updated from the prior parameters of the Normal-Inverse-Gamma distribution, similar to the aforementioned Case 2, given by (4.30) with $i = 1$. Then, the posterior on μ and ν^2 given $l_0 = 0$ and traffic sample x_0 at time step 0, represented by $\Pr\left(\mu, \nu^2 \middle| l_0 = 0, x_0\right)$, is given by (4.27) with $i = 1$. The posterior predictive model for time step $i = 1$, represented by $\Pr\left(x_1 \middle| l_0 = 0, x_0\right)$, is given by (4.28) with $i = 1$.

Let $\mathbf{s}_i^{l'} = [\mu_{i-1}^{(l)}, \kappa_{i-1}^{(l)}, \alpha_{i-1}^{(l)}, \beta_{i-1}^{(l)}]$ be a sufficient statistics vector corresponding to $l_i = l'$ and $\boldsymbol{x}_i^{(l)}$ at time step i. Then, the online update on the sufficient statistics vectors is illustrated in Fig 4.8 for time steps up to $i = 3$.

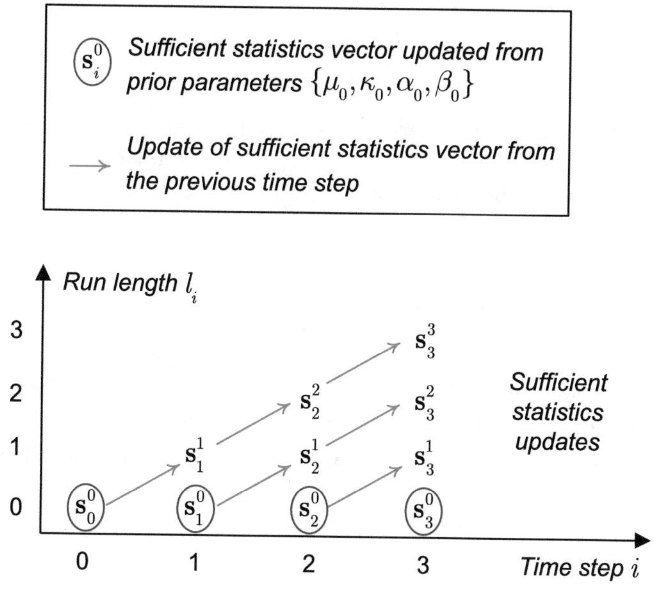

Fig. 4.8 An illustration of sufficient statistics update

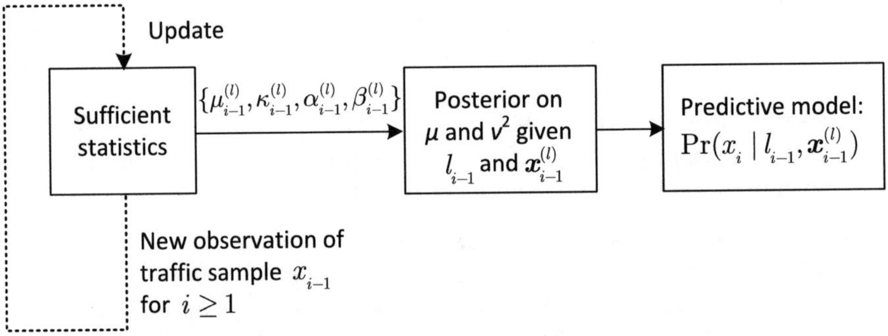

Fig. 4.9 A diagram for online updating of the predictive model

With the updated sufficient statistics, a diagram for the online updating of the predictive model $\Pr\left(x_i \middle| l_{i-1}, \boldsymbol{x}_{i-1}^{(l)}\right)$ for $i \geq 1$ is shown in Fig. 4.9.

From Joint Distribution to Posterior Distribution

With the conditional prior on run length and the predictive model, joint probability $\Pr(l_i, \boldsymbol{x}_i)$ at time step i can be calculated from joint probability $\Pr(l_{i-1}, \boldsymbol{x}_{i-1})$ at time step $i-1$, based on (4.22). Then, the posterior distribution of run length given the observed time series $\boldsymbol{x}_i$ at time step t, i.e., $\Pr(l_i|\boldsymbol{x}_i)$, is given by

$$\Pr(l_i = i'|\boldsymbol{x}_i) = \frac{\Pr(l_i = i', \boldsymbol{x}_i)}{\Pr(\boldsymbol{x}_i)}, \quad \forall i' = 0, 1, \cdots, i \tag{4.31}$$

where $\Pr(\boldsymbol{x}_i)$ is unknown. As mentioned, the joint probability represents a relative likelihood with the initialization $\Pr(l_0 = 0, x_0) = 1$, for any observed value of x_0. Therefore, the posterior distribution of run length, $\Pr(l_i|\boldsymbol{x}_i)$, can be obtained through normalization, given by

$$\Pr(l_i = i'|\boldsymbol{x}_i) = \frac{\Pr(l_i = i', \boldsymbol{x}_i)}{\sum_{l_i=0}^{i} \Pr(l_i, \boldsymbol{x}_i)}, \quad \forall i' = 0, 1, \cdots, i. \tag{4.32}$$

The posterior run length distribution at each time step is the output of the BOCPD algorithm, which has a linear space and time complexity per time-step in the number of medium-timescale traffic samples after the previously detected change point [6].

4.2.1.3 From Stochastic Run Length Distribution to Deterministic Change Points

With the posterior distribution of run length $\Pr(l_i|\boldsymbol{x}_i)$ at time step i, define the most probable run length given the observed time series $\boldsymbol{x}_i$ at time step i as

$$\hat{l}_i = \text{argmax}_{l_i=\{0,\cdots,i\}}\Pr(l_i|\boldsymbol{x}_i). \tag{4.33}$$

Corresponding to $l_i = \hat{l}_i$ and $\boldsymbol{x}_i^{(\hat{l}_i)}$, we have sufficient statistics $\{\mu_i^{(\hat{l}_i)}, \kappa_i^{(\hat{l}_i)}, \alpha_i^{(\hat{l}_i)}, \beta_i^{(\hat{l}_i)}\}$.

Then, the student-t predictive model corresponding to $\hat{l}_i$ has mean $\mu_i^{(\hat{l}_i)}$ and standard deviation $\nu_i^{(\hat{l}_i)}$ given by

$$\nu_i^{(\hat{l}_i)} = \sqrt{\frac{\beta_i^{(\hat{l}_i)}\left(\kappa_i^{(\hat{l}_i)}+1\right)}{\kappa_i^{(\hat{l}_i)}\left(\alpha_i^{(\hat{l}_i)}-1\right)}}. \tag{4.34}$$

The mean and standard deviation of the student-t predictive model corresponding to the most probable run length at time step i, i.e., $\hat{l}_i$, is seen as the most probable mean and standard deviation of the non-stationary medium-timescale time series at time step i.

At each time step i, both the most probable run length and the most probable mean and variance are obtained, based on which time step i is identified as a change point if the following two conditions are satisfied.

- First, the gap between the most probable run lengths at time steps $(i-1)$ and i, i.e., $\hat{l}_{i-1}$ and $\hat{l}_i$, is larger than a predefined threshold Υ_l, given by

$$\hat{l}_{i-1} - \hat{l}_i > \Upsilon_l. \tag{4.35}$$

  Ideally, if time step i indicates a new change point, the run length at time step i deterministically drops to zero. However, in practice, there are possibly multiple non-zero masses in the posterior run length distribution at time step i. The masses for small values of run length including zero are relatively larger than those for large values, and the most probable run length is more likely to be a small value. Hence, we use changes in the most probable run length as the indicator for change point. If the most probable run length suddenly drops to a smaller value than the previous time step, it is highly possible to have a new change point. We use a threshold Υ_l to deterministically determine whether a new change point is generated at each time step. The selection of the threshold is non-trivial. If the threshold is too large, it is possible to delay the detection of some change points; if the threshold is too small, the statistical detection method is sensitive to randomness in the traffic samples. In practice, the selection should be based on experience;

- Second, the normalized absolute difference between the most probable mean plus standard deviation at time steps i and $(i-1)$ is beyond a predefined threshold Υ_d, given by

$$\frac{\left|\left(\mu_i^{(\hat{l}_i)} + \nu_i^{(\hat{l}_i)}\right) - \left(\mu_{i-1}^{(\hat{l}_{i-1})} + \nu_{i-1}^{(\hat{l}_{i-1})}\right)\right|}{\mu_{i-1}^{(\hat{l}_{i-1})} + \nu_{i-1}^{(\hat{l}_{i-1})}} > \Upsilon_d. \tag{4.36}$$

  If time step i indicates a new change point, the traffic sample at time step i belongs to a different traffic segment from the traffic sample in the previous time step $i-1$. The two traffic segments have different statistics such as mean and variance. Hence, we use the relative changes in the combination the mean and stand deviation as a indicator for change points. Similarly, the threshold Υ_d should be carefully selected.

With the initial change point $C_M(0) = 0$ and the change point detection, the k-th detected change point, denoted by $\hat{C}_M(k)$, is seen as the beginning of a new traffic segment. Here, $\hat{C}_M(k)$ is an estimated value of the real change point $C_M(k)$, i.e., the index of the first medium-timescale traffic sample in the k-th stationary traffic segment.

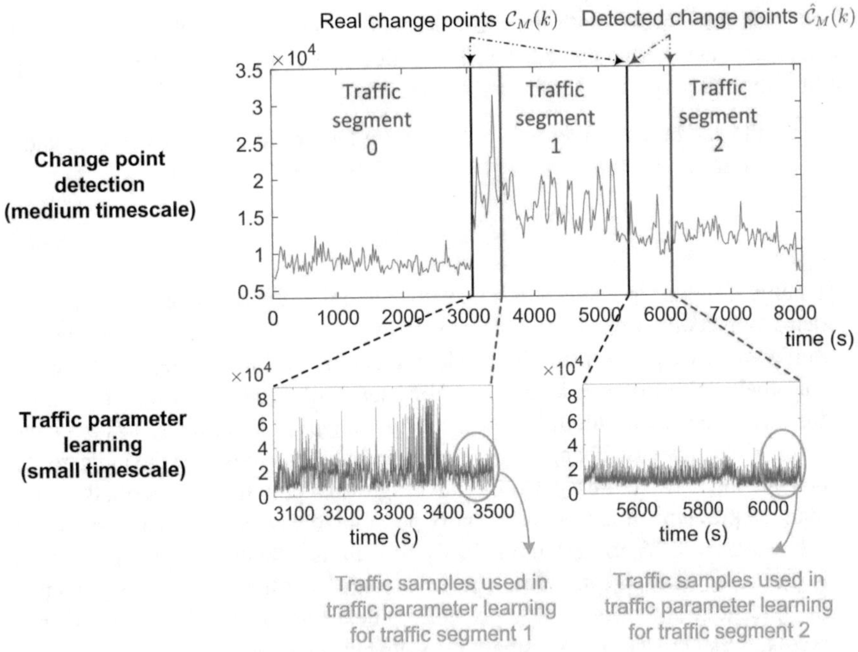

Fig. 4.10 An illustration of *look-back* traffic parameter learning

Due to the stochastic nature of the BOCPD algorithm, there is usually a latency between $C_M(k)$ and $\hat{C}_M(k)$ with $\hat{C}_M(k) > C_M(k)$, as illustrated in Fig. 4.10. In the figure, the real and detected change points in a non-stationary traffic time series in medium timescale are indicated by the black and red vertical lines, respectively. The latency between the real and detected change points cannot be avoided, since it is inherent to the BOCPD algorithm. We exploit the latency for a *look-back* traffic parameter learning.

4.2.2 *Traffic Parameter Learning*

4.2.2.1 A Look-Back Scheme for Traffic Sample Collection

As illustrated in Fig. 4.10, the small-timescale traffic samples between the real change point $C_M(k)$ and the detected change point $\hat{C}_M(k)$ all belong to traffic segment k. For traffic segment 1, the delay between the real and detected change points is around 420 s, corresponding to 21 and 4200 traffic samples in medium and small timescales, respectively, with $T_M = 20$ s and $T_S = 100$ ms. It demonstrates that about 21 more time steps are required for the detection of change point $\hat{C}_M(1)$ in medium timescale. The 4200 small-timescale traffic samples can be used to learn the traffic parameters of traffic segment 1. However, we only use a most recent subset of the 4200 small-timescale traffic samples for traffic parameter learning in practice, as indicated by a green circle in Fig. 4.10, due to two reasons. First, the real change points are unknown, and it is difficult to know exactly how many small-timescale traffic samples before the detected change point belong to the same traffic segment with the traffic samples after the change point. Second, the computation complexity is high if too many traffic samples are used in traffic parameter learning.

Next, we formally define the subset of small-timescale traffic samples used to learn the traffic parameters of traffic segment k. Let i_0 be a small integer[2] such that $(i_0 - 1)$ medium-timescale traffic samples before the $\hat{C}_M(k)$-th one belong to the k-th stationary traffic segment. The $\frac{i_0 T_M}{T_S}$ small-timescale traffic samples in the same time duration with the i_0 medium-timescale traffic samples including the $\hat{C}_M(k)$-th one are used to learn the traffic parameters of the k-th stationary traffic segment. The mapping between the i_0 medium-timescale traffic samples and the $\frac{i_0 T_M}{T_S}$ small-timescale traffic samples is given by

$$
\begin{aligned}
&\boldsymbol{x}_M[(\hat{C}_M(k) - i_0 + 1) : \hat{C}_M(k)] \\
\Rightarrow &\boldsymbol{x}_S\left[\frac{(\hat{C}_M(k) - i_0 + 1)T_M}{T_S} : \left(\frac{(\hat{C}_M(k) + 1)T_M}{T_S} - 1\right)\right]
\end{aligned}
$$

[2] The value of i_0 should be smaller than the minimum value of the most probable run lengths at any detected change points.

$$=\boldsymbol{x}_S\left[\hat{C}_S(k):\left(\hat{C}_S(k)+\frac{i_0T_M}{T_S}-1\right)\right] \tag{4.37}$$

where $\hat{C}_S(k) = \frac{(\hat{C}_M(k)-i_0+1)T_M}{T_S}$ is the estimated k-th change point in small timescale. Since the traffic samples used for traffic parameter learning is before and at the detected change point of a target traffic segment, we refer to our traffic parameter leaning scheme as a *look-back* scheme. Compared with a *look-ahead* counterpart which uses traffic samples after the detected change point, the *look-back* scheme avoids another latency for collecting sufficient traffic samples.

4.2.2.2 Fractional Brownian Motion Traffic Model

As discussed, for the k-th stationary traffic segment, the small-timescale traffic time series $\boldsymbol{x}_S\left[\hat{C}_S(k):\left(\hat{C}_S(k)+\frac{i_0T_M}{T_S}-1\right)\right]$ is used for traffic parameter learning.

We consider a modified shifted discrete time line, $\tilde{t}$, with $\tilde{t}=t-\hat{C}_S(k)$, for the k-th stationary traffic segment. Correspondingly, we have $\tilde{x}_S(\tilde{t}) = x_S(t-\hat{C}_S(k))$ for $\hat{C}_S(k) \le t \le \left(\hat{C}_S(k)+\frac{i_0T_M}{T_S}-1\right)$, which represents the number of packets arrived in the $\tilde{t}$-th modified shifted small time interval. Then, the traffic time series in the shifted small-timescale time line used for traffic parameter learning is $\tilde{\boldsymbol{x}}_S\left[0:\left(\frac{i_0T_M}{T_S}-1\right)\right]$.

The cumulative number of packet arrivals in the k-th stationary traffic segment before shifted time point $\tilde{t}$, is given by

$$\tilde{\mathsf{A}}_k\left(\tilde{t}\right)=\begin{cases}\sum_{\tilde{t}'=1}^{\tilde{t}}\tilde{x}_S(\tilde{t}'-1), & 1\le\frac{i_0T_M}{T_S}-1\\ 0, & \tilde{t}=0.\end{cases} \tag{4.38}$$

We model $\tilde{\mathsf{A}}_k\left(\tilde{t}\right)$ as a general fBm process with traffic parameters $\{\lambda\,(k), \sigma\,(k), H\,(k)\}$, which are the fBm traffic parameters of the k-th stationary traffic segment.

4.2.2.3 Gaussian Process Regression

With the fBm traffic model, $\tilde{\mathsf{A}}_k\left(\tilde{t}\right)$ is a Gaussian process with mean function $\lambda\,(k)\,\tilde{t}$ and fBm covariance function $\psi_k\left(\tilde{t}_1,\tilde{t}_2\right)$ given by

$$\psi_k\left(\tilde{t}_1,\tilde{t}_2\right)=\frac{\sigma^2(k)}{2}\left(\tilde{t}_1^{2H(k)}+\tilde{t}_2^{2H(k)}-|\tilde{t}_1-\tilde{t}_2|^{2H(k)}\right). \tag{4.39}$$

Accordingly, $\tilde{\mathsf{A}}_k\left(\tilde{t}\right)$ is represented as

$$\tilde{\mathsf{A}}_k\left(\tilde{t}\right) \sim \mathcal{GP}\left(\lambda\left(k\right)\tilde{t},\ \psi_k\left(\tilde{t}_1, \tilde{t}_2\right)\right). \tag{4.40}$$

The fBm traffic parameters $\{\lambda(k), \sigma(k), H(k)\}$ can be inferred based on Gaussian process regression, in which $\{\lambda(k), \sigma(k), H(k)\}$ are referred to as hyper-parameters [3, 9].

To train a Gaussian process regression model, let $\boldsymbol{t}_k = \left[0, 1, \cdots, \left(\frac{i_0 T_M}{T_S} - 1\right)\right]$ be training inputs and $\mathbf{A}_k = \left[\tilde{\mathsf{A}}_k(0), \tilde{\mathsf{A}}_k(1), \cdots, \tilde{\mathsf{A}}_k\left(\frac{i_0 T_M}{T_S} - 1\right)\right]$ be training outputs. Then, we have the following joint Gaussian distribution

$$\mathbf{A}_k \sim \mathcal{N}\left(\lambda\left(k\right)\boldsymbol{t}_k, \Psi_k\right) \tag{4.41}$$

where Ψ_k is a $\frac{i_0 T_M}{T_S}$-by-$\frac{i_0 T_M}{T_S}$ covariance matrix with $\Psi\left(\mathtt{j}, \mathtt{j}'\right) = \psi_k\left(\mathtt{j}, \mathtt{j}'\right)$, represented as

$$\Psi_k = \begin{bmatrix} \psi_k(0,0) & \psi_k(0,1) & \cdots & \psi_k\left(0, \frac{i_0 T_M}{T_S} - 1\right) \\ \psi_k(1,0) & \psi_k(1,1) & \cdots & \psi_k\left(1, \frac{i_0 T_M}{T_S} - 1\right) \\ \vdots & \vdots & \ddots & \vdots \\ \psi_k\left(\frac{i_0 T_M}{T_S} - 1, 0\right) & \psi_k\left(\frac{i_0 T_M}{T_S} - 1, 1\right) & \cdots & \psi_k\left(\frac{i_0 T_M}{T_S} - 1, \frac{i_0 T_M}{T_S} - 1\right) \end{bmatrix}. \tag{4.42}$$

The Gaussian process regression model is trained, i.e., the hyper-parameters are learned, by maximizing the following log-marginal likelihood function with a gradient optimizer

$$\log \Pr\left(\mathbf{A}_k \mid \boldsymbol{t}_k; \{\lambda(k), \sigma(k), H(k)\}\right) =$$
$$-\frac{1}{2}\left[\left(\mathbf{A}_k - \lambda(k)\boldsymbol{t}_k\right)^T \Psi_k^{-1}\left(\mathbf{A}_k - \lambda(k)\boldsymbol{t}_k\right) + \log|\Psi_k| + \frac{i_0 T_M}{T_S}\log 2\pi\right]. \tag{4.43}$$

After the training, the fBm traffic parameter leaning for stationary traffic segment k is done.

The Gaussian process regression with $\frac{i_0 T_M}{T_S}$ traffic samples has $O\left(\frac{i_0 T_M}{T_S}^3\right)$ time complexity and $O\left(\frac{i_0 T_M}{T_S}^2\right)$ space complexity due to the inversion of a covariance matrix in (4.43) [10]. Such a complexity is feasible on a desktop computer for dataset sizes up to a few thousands. There are sparse approximation algorithms to reduce the complexity of Gaussian process regression [10].

To evaluate the learning accuracy, one-step-ahead predictions for t_0 subsequent small time intervals, i.e., $\frac{i_0 T_M}{T_S} \leq \tilde{t} \leq \frac{i_0 T_M}{T_S} + t_0 - 1$, are performed using the trained Gaussian process regression model. The one-step-prediction for small time interval $\tilde{t}$ is to predict $\tilde{\mathsf{A}}_k\left(\tilde{t}\right)$, given a set of observed data $\mathcal{D} = (\boldsymbol{t}, \mathbf{A})$ at the previous time interval $\tilde{t} - 1$, with $\boldsymbol{t} = [0, 1, \cdots, \tilde{t} - 1]$ and $\mathbf{A} = [\tilde{\mathsf{A}}_k(0), \tilde{\mathsf{A}}_k(1), \cdots, \tilde{\mathsf{A}}_k\left(\tilde{t} - 1\right)]$.

For small time interval $\tilde{t}$, the trained Gaussian process regression model gives a Gaussian posterior distribution of $\tilde{\mathsf{A}}_k(\tilde{t})$ conditioned on $\tilde{t}$ and $\mathcal{D}$, as

$$\Pr\left(\tilde{\mathsf{A}}_k(\tilde{t})\,|\tilde{t},\mathcal{D}\right) \sim \mathcal{N}\left(\mu_{\mathcal{GP};k}(\tilde{t}),\sigma^2_{\mathcal{GP};k}(\tilde{t})\right) \tag{4.44}$$

with

$$\begin{cases} \mu_{\mathcal{GP};k}(\tilde{t}) = & \boldsymbol{\psi}_k(\boldsymbol{t},\tilde{t})^T(\Psi)^{-1}\mathbf{A} \\ \sigma^2_{\mathcal{GP};k}(\tilde{t}) = \psi_k(\tilde{t},\tilde{t}) - \boldsymbol{\psi}_k(\boldsymbol{t},\tilde{t})^T(\Psi)^{-1}\boldsymbol{\psi}_k(\boldsymbol{t},\tilde{t}) \end{cases} \tag{4.45}$$

where $\boldsymbol{\psi}_k(\boldsymbol{t},\tilde{t})$ is a $\tilde{t}$-by-1 vector with the $\mathtt{j}$-th component equal $\psi_k(\mathtt{j},\tilde{t})$, given by

$$\boldsymbol{\psi}_k(\boldsymbol{t},\tilde{t}) = \left[\psi_k(0,\tilde{t}),\psi_k(1,\tilde{t}),\cdots,\psi_k(\tilde{t}-1,\tilde{t})\right] \tag{4.46}$$

and Ψ is a $\tilde{t}$-by-$\tilde{t}$ covariance matrix with $\Psi(\mathtt{j},\mathtt{j}') = \psi_k(\mathtt{j},\mathtt{j}')$, represented as

$$\Psi = \begin{bmatrix} \psi_k(0,0) & \psi_k(0,1) & \cdots & \psi_k(0,\tilde{t}-1) \\ \psi_k(1,0) & \psi_k(1,1) & \cdots & \psi_k(1,\tilde{t}-1) \\ \vdots & \vdots & \ddots & \vdots \\ \psi_k(\tilde{t}-1,0) & \psi_k(\tilde{t}-1,1) & \cdots & \psi_k(\tilde{t}-1,\tilde{t}-1) \end{bmatrix}. \tag{4.47}$$

Based on (4.45), the mean, $\mu_{\mathcal{GP};k}(\tilde{t})$, is taken as a point estimate for the prediction output, and the variance, $\sigma^2_{\mathcal{GP};k}(\tilde{t})$, provides an uncertainty measure for the point estimate. Then, the traffic sample in time interval $\tilde{t}$, i.e., $\tilde{x}_S(\tilde{t})$, is predicted as

$$\hat{\tilde{x}}_S(\tilde{t}) = \mu_{\mathcal{GP};k}(\tilde{t}) - \tilde{\mathsf{A}}_k(\tilde{t}-1). \tag{4.48}$$

The prediction error of the t_0 traffic samples in the k-th stationary traffic segment, E_k, is defined as the normalized root-mean-squared deviation between the t_0 predicted traffic samples and the corresponding ground truth, given by

$$\mathrm{E}_k = \frac{\sqrt{\sum_{\tilde{t}=\frac{i_0 T_M}{T_S}}^{\frac{i_0 T_M}{T_S}+t_0-1}\left(\hat{\tilde{x}}_S(\tilde{t}) - \tilde{x}_S(\tilde{t})\right)^2}}{\left(x_S^{max} - x_S^{min}\right)\sqrt{t_0}}. \tag{4.49}$$

The normalization constant is the scale of small-timescale traffic samples, i.e., $\left(x_S^{max} - x_S^{min}\right)$. A smaller E_k value indicates a higher learning accuracy for traffic parameters.

4.2.3 Resource Demand Prediction for a Stationary Traffic Segment

Once change point $\hat{C}_M(k)$ is detected, the fBm traffic parameters of the upcoming k-th stationary traffic segment, i.e., $\{\lambda(k), \sigma(k), H(k)\}$, are learned from the most recent $\frac{i_0 T_M}{T_S}$ historical small-timescale traffic samples based on Gaussian process regression. Such a *look-back* traffic parameter learning scheme enables resource demand prediction for the upcoming k-th stationary traffic segment.

Let $\mathsf{R}(k)$ be the minimum computing resource demand (in packet/s)[3] of the k-th stationary traffic segment at the considered VNF. With the fBm resource provisioning model given in Sect. 4.1.3, $\mathsf{R}(k)$ can be calculated from the learned fBm traffic parameters, $\{\lambda(k), \sigma(k), H(k)\}$, the probabilistic delay requirement, $\Pr(\mathsf{d} > \mathsf{D}) \leq \varepsilon$, and the small time interval length, T_S.

4.3 Dynamic VNF Resource Scaling and Migration Framework

Figure 4.11 shows a diagram for a learning-based VNF resource scaling and migration framework with non-stationary traffic.

In the diagram, the change point detection driven and traffic parameter learning based resource demand prediction scheme for non-stationary traffic, as introduced in Sect. 4.2,is summarized in a resource demand prediction module. With the non-stationary traffic time series in medium timescale, the BOCPD algorithm locates the prior-unknown change points of the non-stationary traffic. The change points determine the boundaries between consecutive stationary traffic segments. With the detected change point $\hat{C}_M(k)$, a subset of small-timescale traffic samples are collected for a *look-back* parameter learning, with which the fBm traffic parameters $\{\lambda k, \sigma(k), H(k)\}$ are learned for the upcoming k-th stationary traffic segment. Then, the resource demand $\mathsf{R}(k)$ is predicted with the fBm resource provisioning model.

The detected change points also determine the boundaries between consecutive decision epochs indexed by k for VNF resource scaling and necessary migrations. That is, the change point detection in the resource demand prediction module provides a triggering signal for dynamic VNF resource scaling and migration decisions. The decision epochs have variable lengths. The length of decision epoch k is equal to $(\hat{C}_M(k+1) - \hat{C}_M(k))T_M$. For resource scaling, the same amount of computing resources as the predicted resource demand $\mathsf{R}(k)$ is allocated to the VNF in epoch k. For VNF migration at decision epoch k, one of the candidate NFV

[3] Here we refer to the VNF packet processing rate as the computing resource demand for simplicity. The CPU computing resource demand in cycle/s can be mapped from the resource demand in packet/s with the processing density model in Chap. 3.

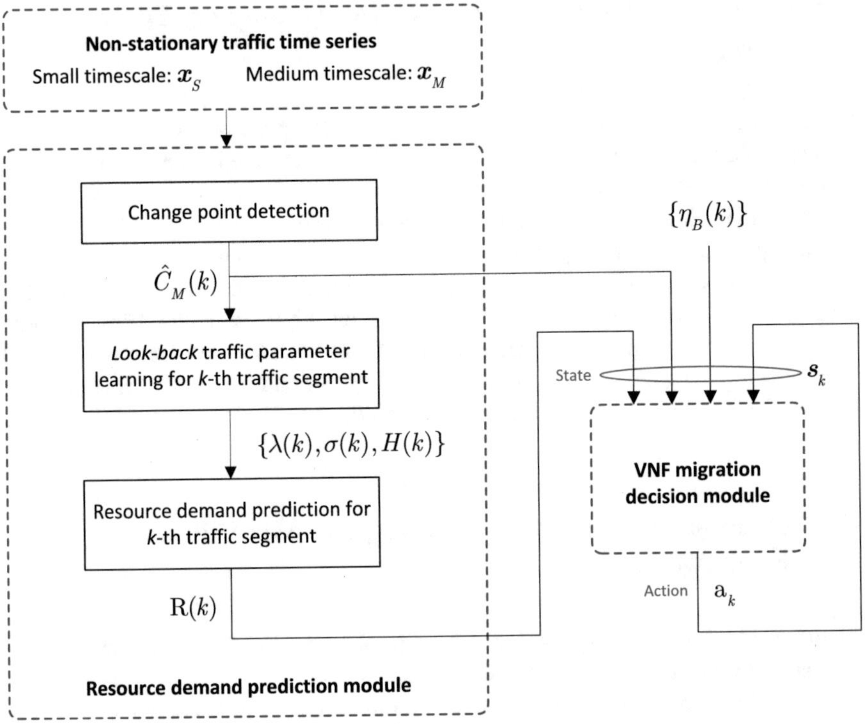

Fig. 4.11 A diagram for the learning-based VNF resource scaling and migration framework with non-stationary traffic

nodes which might be different from the current placed NFV node is selected as the new supporting NFV node for the VNF, based on a separate VNF migration decision module in the diagram. More details on VNF migration decision are given in Sect. 4.4.

4.4 Dynamic VNF Migration Decision

As demonstrated in Chap. 3, there is a trade-off between load balancing and migration cost in the VNF migration decision. Hence, the decision for epoch k should be made with the consideration of migration cost from epoch $k-1$ to epoch k and the background resource loading status in the considered network segment.

For a one-shot VNF migration problem, it should determine the re-mapping between VNF and candidate NFV nodes subject to computing resource capacity constraints at the NFV nodes, to minimize the maximum resource utilization factor among NFV nodes for load balancing, while minimizing the migration cost.

However, to address the trade-off between load balancing and migration cost in the long run, the dynamic VNF migration decisions across different decision epochs is modeled a Markov decision process.

4.4.1 Markov Decision Process

A Markov decision process (MDP) is formally defined as a tuple $< \mathcal{S}, \mathcal{A}, P, R >$, where $\mathcal{S}$ and $\mathcal{A}$ are the state spaces, $\mathcal{P} : \mathcal{S} \times \mathcal{A} \times \mathcal{S} \to [0, 1]$ is the transition probability, and R is the reward model. Specifically, $\Pr(s'|s, a)$ is the transition probability from state $s \in \mathcal{S}$ to state $s' \in \mathcal{S}$ after taking action $a \in \mathcal{A}$, and $R(r|s, a)$ is the probability of receiving reward r for taking action $a \in \mathcal{A}$ in state $s \in \mathcal{S}$. For a deterministic reward model, we have a reward function $R : \mathcal{S} \times \mathcal{A} \to \mathbb{R}$ with $r = R(s, a)$. A stochastic policy, $\pi(a|s)$, specifies the probability of taking action $a \in \mathcal{A}$ at state $s \in \mathcal{S}$. A deterministic policy, $a = \pi(s)$, deterministically maps state $s \in \mathcal{S}$ to action $a \in \mathcal{A}$.

For a deterministic policy with a deterministic reward function, a state-value function for policy π at state s, denoted by $V^{\pi}(s)$, is defined as the expected discounted cumulative reward when starting at state s and following policy π thereafter, given by

$$V^{\pi}(s) = \mathbb{E}_{\pi}\left[\sum_{k=0}^{\infty} \gamma^k r_k | s_0 = s\right] \tag{4.50}$$

where $\gamma \in (0, 1]$ is the discount factor, and s_0 denotes the state at epoch $k = 0$.

Similarly, a state-action value function $Q^{\pi}(s, a)$, denoting the value of taking action a at states under the policy π, is defined as

$$Q^{\pi}(s, a) = \mathbb{E}_{\pi}\left[\sum_{k=0}^{\infty} \gamma^k r_k | s_0 = s, a_0 = a\right] \tag{4.51}$$

where a_0 denotes the action at epoch $k = 0$

The goal to find the optimal policy, $\pi^* = \arg\max_{\pi} Q^{\pi}(s, a)$, which maximizes the state-action value for all s and a. The optimal state-action value function $Q^*(s, a)$ satisfies the Bellman equation

$$Q^*(s, a) = \mathbb{E}_{\pi^*}\left[R(s, a) + \gamma \max_{a' \in \mathcal{A}} Q^*(s', a')\right]. \tag{4.52}$$

With the transition probability $\mathcal{P}$, the optimal state-action value function $Q^*(s, a)$ can be found based on the Bellman equation with dynamic programming. Starting from any $Q_0(s, a)$ at epoch 0, $Q_k(s, a)$ converges to $Q^*(s, a)$ at $k \to \infty$ by iteratively applying the Bellman equation.

We model the dynamic VNF migration problem as an MDP. Next, we define the state, action, and reward for dynamic VNF migration. The state and action are also shown in Fig. 4.11. Here, as part of the state is the action of the previous epoch in our VNF migration problem, we first introduce the action.

4.4.1.1 Action

Let $\{\mathrm{a}_k^n, n \in \mathcal{N}_C\}$ be a binary variable set, with $\mathrm{a}_k^n = 1$ if the VNF is placed at NFV node $n \in \mathcal{N}_C$ during the k-th decision epoch, and $\mathrm{a}_k^n = 0$ otherwise. Let a_k $(0 \leq \mathrm{a}_k \leq |\mathcal{N}_C| - 1)$ be an integer denoting the VNF location during decision epoch k, with $\mathrm{a}_k = n$ if the VNF is placed at NFV node $n \in \mathcal{N}_C$. The relationship between $\{\mathrm{a}_k^n\}$ and a_k is given by

$$\mathrm{a}_k^n = \begin{cases} 1, \text{ if } \mathrm{a}_k = n \\ 0, \text{ otherwise.} \end{cases} \tag{4.53}$$

The action at decision epoch k is the new VNF location indicated by a_k, which is associated with a discrete action space with dimensionality $|\mathcal{N}_C|$. The discrete action space is $\mathcal{A} = \{0, 1, \cdots, |\mathcal{N}_C| - 1\}$. We use a_k instead of $\{\mathrm{a}_k^n\}$ as the action to limit the dimensionality of action space.

4.4.1.2 State

Let $\boldsymbol{s}_k$ denote the state at decision epoch k. As mentioned, the action of the previous epoch, i.e., a_{k-1}, is part of the state, to represent the initial VNF placement status at the beginning of the current epoch k. Whether a VNF migration happens or not in epoch k is uniquely determined by the initial VNF placement state a_{k-1} and the action a_k.

The predicted resource demand of the VNF during epoch k, i.e., $\mathsf{R}(k)$, provided by the resource demand prediction module, is another element in state $\boldsymbol{s}_k$.

For decision epochs with variable lengths, the starting time (in hour) of decision epoch k is also an element in state $\boldsymbol{s}_k$, which is provided by the change point detection algorithm in the resource demand prediction module. Let $\hat{C}(k)$ $(0 \leq \hat{C}(k) < 24)$ be a real number representing the starting time (in hour) of the k-th change point. It can be calculated from the detected change point $\hat{C}_M(k)$, given by

$$\hat{C}(k) = \frac{\hat{C}_M(k) T_M}{3600} \mod 24 \tag{4.54}$$

where the modulo operation limits $\hat{C}(k)$ in $[0, 24)$.

Let $\eta_n^B(k)$ denote the average background resource utilization factor at NFV node $n \in \mathcal{N}_C$ during decision epoch k, which is the average ratio between the amount of computing resources (in packet/s) allocated to background traffic at NFV node $n \in \mathcal{N}_C$ during decision epoch k and the computing resource capacity R_n (in packet/s) of NFV node $n \in \mathcal{N}_C$. The average background resource utilization factor at NFV node $n \in \mathcal{N}_C$ changes dynamically across different decision epochs. The average background resource utilization factors of all the candidate NFV nodes during decision epoch k, i.e., $\{\eta_n^B(k)\}$, are also part of the state, to represent the current loading status in the considered network segment.

In summary, the state at decision epoch k is composed of four parts, and represented as $\boldsymbol{s}_k = [\mathrm{a}_{k-1}, \mathsf{R}(k), \hat{C}(k), \{\eta_n^B(k)\}]$.

4.4.1.3 Reward

In the considered network scenario, if the VNF is placed at NFV node $n \in \mathcal{N}_C$, the overall traffic at NFV node $n \in \mathcal{N}_C$ is an aggregation of both the background traffic and the traffic from the considered VNF; otherwise, NFV node $n \in \mathcal{N}_C$ is loaded with only the background traffic. Let $\eta_n(k)$ denote the average resource utilization factor at NFV node $n \in \mathcal{N}_C$ during decision epoch k, representing the ratio between the average amount of computing resources (in packet/s) allocated to the overall traffic at NFV node $n \in \mathcal{N}_C$ during decision epoch k and R_n. For $\eta_n(k)$, it depends on $\eta_n^B(k)$ and $\mathsf{R}(k)$ in state $\boldsymbol{s}_k$ and the VNF placement action a_k, given by

$$\eta_n(k) = \eta_n^B(k) + \frac{\mathrm{a}_k^n \mathsf{R}(k)}{\mathsf{R}_n}. \tag{4.55}$$

Load Balancing With $\{\eta_n(k), \forall n \in \mathcal{N}_C\}$, the cost for imbalanced loading during decision epoch k, denoted by $c_k^{(1)}$, is defined as the maximum average resource utilization factor among all NFV nodes in $\mathcal{N}_C$, given by

$$c_k^{(1)} = \max_{n \in \mathcal{N}_C} \eta_n(k). \tag{4.56}$$

Minimizing $c_k^{(1)}$ achieves the maximum load balancing among all the candidate NFV nodes.

Migration Cost Assume that each VNF migration incurs the same migration cost. Then, we can use the total number of migrations to denote the total migration cost. Let $c_k^{(2)}$ denote the total migration cost in epoch k, which is a binary variable as we consider migration for a single VNF in a local network segment. We have

$$c_k^{(2)} = \begin{cases} \sum_{n \in \mathcal{N}_C} \sum_{n' \in \mathcal{N}_C \setminus n} \mathrm{a}_{k-1}^n \mathrm{a}_k^{n'}, \text{ if } k > 0 \\ 0, \text{ if } k = 0 \end{cases} \tag{4.57}$$

where a_{k-1}^{n} is a known value given state s_k. We have $c_k^{(2)} = 1$ for $k > 0$ if the placed NFV node of the VNF changes from $\forall n \in \mathcal{N}_C$ in epoch $(k-1)$ to $\forall n' \in \mathcal{N}_C \backslash n$ epoch k, and $c_k^{(2)} = 0$ otherwise.

Resource Overloading Penalty In a one-shot VNF migration problem for epoch k, a total cost integrating both load balancing and migration cost should be minimized, subject to computing resource capacity constraints at each NFV node, i.e., $\eta_n(k) \leq 1, \forall n \in \mathcal{N}_C$. In an unconstrained MDP, no explicit resource capacity constraints are included, so a bad action might result in resource overloading at the placed NFV node, which should be penalized. Let $\mathrm{f}_k^{(\mathrm{P})}$ be a binary flag indicating whether there is penalty due to resource overloading or not, and $c^{(\mathrm{P})}$ be a constant representing the level of penalty. Assume that resource overloading is only due to improper VNF placement, i.e., the background traffic does not overload the NFV nodes ($\eta_n^B(k) < 1$). Then, the penalty flag is defined as

$$\mathrm{f}_k^{(\mathrm{P})} = \begin{cases} 1, & c_k^{(1)} > \eta_U \\ 0, & \text{otherwise} \end{cases} \tag{4.58}$$

where η_U ($0 < \eta_U \leq 1$) is an upper bound for the maximum resource utilization factor without penalty. As the resource overloading penalty due to resource capacity constraint violation cannot be completely avoided in a learning-based solution with random exploration, we select η_U as a number close to but smaller than 1, e.g., $\eta_U = 0.95$, to penalize utilization factors close to 1. Here, we assume that there is always a feasible NFV node for VNF placement without resource overloading. In practice, the predicted resource demand can be very large if the historical traffic shows a significant increasing trend. In this case, it is possible that there is no feasible VNF placement without resource overloading. A potential solution is to throttle the traffic when resource overloading is foreseen to happen. For simplicity, we do not consider traffic throttling here.

With the models for the cost of imbalanced loading, the migration cost, and the resource overloading penalty, we define a deterministic reward function $r_k = R(s_k, a_k)$, given by

$$\mathrm{r}_k = -\left[\omega^{(\mathrm{C})} c_k^{(1)} + \left(1 - \omega^{(\mathrm{C})}\right) c_k^{(2)} + c^{(\mathrm{P})} \mathrm{f}_k^{(\mathrm{P})}\right] \tag{4.59}$$

where $\omega^{(\mathrm{C})}$ is a weighting factor in (0, 1).

4.4.2 Deep Reinforcement Learning

In the dynamic environment of VNF migration, the transition probability $\mathcal{P}$ is unavailable. Therefore, we solve the MDP by a reinforcement learning (RL) approach. We consider an episodic task for the VNF migration decisions, in which the RL agent interacts with the dynamic VNF migration environment in a sequence of episodes. An episode corresponds to a time duration such as 1 day, 1 week, or 1 month, depending on periodic features of traffic dynamics. Each episode consists of a finite number of learning steps. In the dynamic VNF migration problem, learning step k corresponds to decision epoch k determined by the change points. Typically, episodic tasks are easier to learn compared with continuous tasks, because each action affects a finite number of rewards subsequently received during the episode.

At the beginning of an episode, the VNF placement is initialized at NFV node $n_0 \in \mathcal{N}_C$. Within an episode, an agent observes state $\boldsymbol{s}_k$ and takes action a_k at the beginning of decision epoch k. At the end of decision epoch k, the agent receives reward r_k, and sees new state $\boldsymbol{s}_{k+1}$. The goal of the RL agent is to find a deterministic policy, $\pi(\boldsymbol{s})$, mapping a state to an action, to maximize the expected discounted cumulative (episodic) reward $\mathbb{E}(\sum_{k=0}^{K-1} \gamma^k \mathrm{r}_k)$, where K is the number of variable-length decision epochs in an episode. In our simulation, we use 1 week as the time duration for an episode. Hence, K varies for different episodes due to the variable-length decision epochs determined by change point detection.

A penalty-aware deep Q-learning algorithm is presented for the RL solution, with a diagram shown in Fig. 4.12 and pseudo code given in Algorithm 4.1. Next, we first introduce the deep Q-learning algorithm, and then introduce the prioritized experience replay with penalty awareness.

4.4.2.1 Deep Q-Learning

In the episodic task setting, $Q(\boldsymbol{s}_k, \mathrm{a}_k)$ denotes the expected discounted cumulative reward from learning step k to the end of an episode, given by

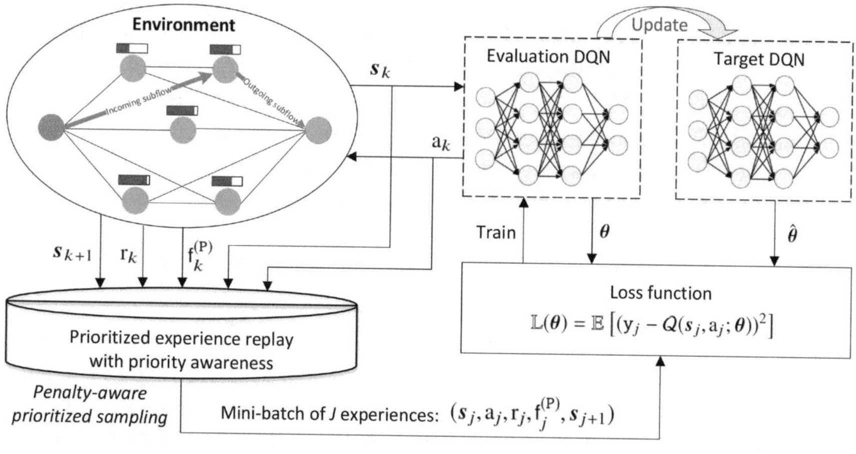

Fig. 4.12 A diagram for the penalty-aware deep Q-learning framework

Algorithm 4.1: Penalty-aware deep Q-learning

Initialize: Evaluation and target DQNs with random weights, set learning parameters as listed in Table 4.1.
for *each episode* **do**
  Initialize VNF placement at NFV node n_0.
  for *each learning step (decision epoch)* **do**
    Observe current state $\boldsymbol{s}_k$, select an action a_k according to the ϵ-greedy policy in (4.61).
    Execute action a_k, collect reward r_k and penalty flag $\mathrm{f}_k^{(\mathrm{P})}$, and see the next state $\boldsymbol{s}_{k+1}$.
    Store transition $(\boldsymbol{s}_k, \mathrm{a}_k, \mathrm{r}_k, \mathrm{f}_k^{(\mathrm{P})}, \boldsymbol{s}_{k+1})$ into replay memory, with initial priority $p_k = \max_{j<k} p_j$.
    for *J iterations* **do**
      Sample a transition $(\boldsymbol{s}_j, \mathrm{a}_j, \mathrm{r}_j, \mathrm{f}_j^{(\mathrm{P})}, \boldsymbol{s}_{j+1})$ with probability $\Pr(j)$.
      Compute importance-sampling weight w_j.
      Compute target value y_j and TD error δ_j.
      Update transition priority p_j.
      Perform a gradient descent, i.e., $\boldsymbol{\theta} \leftarrow \boldsymbol{\theta} + \alpha_Q \left(\mathrm{w}_j \delta_j\right) \nabla_{\boldsymbol{\theta}} Q(\boldsymbol{s}_j, \mathrm{a}_j)$.
    Decrease the exploration probability ϵ_2 by a step $\epsilon_{2,\Delta}$, if $\epsilon_2 > \epsilon_2^{min}$.
    Every $K_{\boldsymbol{\theta}}$ steps, set $\hat{\boldsymbol{\theta}} = \boldsymbol{\theta}$.
Output: Trained evaluation and target DQNs.

$$Q(\boldsymbol{s}_k, \mathrm{a}_k) = \mathbb{E}\left[\sum_{k'=k}^{K-1} \gamma^{k'-k} r_{k'} | \boldsymbol{s}_k, \mathrm{a}_k\right]. \tag{4.60}$$

The MDP for dynamic VNF migration is featured by a high-dimensional combinational state space and a low-dimensional discrete action space. To tackle the curse of dimensionality, deep Q-learning adopts two deep Q-networks (DQNs) with the same neural network structure as Q function approximators, i.e., evaluation DQN (Q) and target DQN ($\hat{Q}$), as illustrated in Fig. 4.12 [11]. The weights of the evaluation DQN (Q) and the target DQN ($\hat{Q}$) are represented as $\boldsymbol{\theta}$ and $\hat{\boldsymbol{\theta}}$, respectively. Every $K_{\boldsymbol{\theta}}$ learning steps, $\hat{\boldsymbol{\theta}}$ is replaced by $\boldsymbol{\theta}$, as illustrated in Fig. 4.12.

The Q-learning is an off-policy algorithm adopting the ϵ-greedy policy

$$\pi(\boldsymbol{s}_k) = \begin{cases} \underset{\mathrm{a}}{\operatorname{argmax}}\, Q(\boldsymbol{s}_k, \mathrm{a}), & \text{with probability } (1-\epsilon_2) \\ \text{random action}, & \text{with probability } \epsilon_2 \end{cases} \tag{4.61}$$

where ϵ_2 is the exploration probability. We use a gradually decreasing ϵ_2 from 1 to a minimum value ϵ_2^{min}, with a step size $\epsilon_{2,\Delta}$, to transit smoothly from exploration to exploitation. The policy in (4.61) is based on evaluation DQN, which gives action a_k, as illustrated in Fig. 4.12.

Traditionally, at epoch k, the evaluation DQN is trained, i.e., $\boldsymbol{\theta}$ is updated, by minimizing a loss function $\mathbb{L}$, given by

$$\mathbb{L}(\boldsymbol{\theta}) = \mathbb{E}\left[(\mathrm{y}_k - Q(\boldsymbol{s}_k, \mathrm{a}_k; \boldsymbol{\theta}))^2\right] \tag{4.62}$$

through gradient descent on $\boldsymbol{\theta}$, where y_k is a target value estimated by target DQN, given by

$$\mathrm{y}_k = \mathrm{r}_k + \gamma \max_{\mathrm{a}} \hat{Q}(\boldsymbol{s}_{k+1}, \mathrm{a}; \hat{\boldsymbol{\theta}}). \tag{4.63}$$

If an episode terminates at the k-th learning step, y_k is set as r_k. A gradient descent on $\boldsymbol{\theta}$ is performed by

$$\boldsymbol{\theta} \leftarrow \boldsymbol{\theta} - \frac{1}{2}\alpha_Q \nabla_{\boldsymbol{\theta}} \mathbb{L}(\boldsymbol{\theta}) = \boldsymbol{\theta} + \alpha_Q \delta_k \nabla_{\boldsymbol{\theta}} Q(\boldsymbol{s}_k, \mathrm{a}_k) \tag{4.64}$$

where α_Q is the learning rate, and δ_k is the temporal-difference (TD) error for transition k, given by

$$\delta_k = \mathrm{y}_k - Q(\boldsymbol{s}_k, \mathrm{a}_k; \boldsymbol{\theta}). \tag{4.65}$$

4.4.2.2 Penalty-Aware Prioritized Experience Replay

Experience Replay

In practice, experience replay is introduced in deep Q-learning for stable convergence [11]. At learning step k, the current transition $(\boldsymbol{s}_k, \mathrm{a}_k, \mathrm{r}_k, \boldsymbol{s}_{k+1})$ is stored in a replay memory. For the training at epoch k, instead of updating $\boldsymbol{\theta}$ by the current transition $(\boldsymbol{s}_k, \mathrm{a}_k, \mathrm{r}_k, \boldsymbol{s}_{k+1})$ at epoch k according to (4.64), $\boldsymbol{\theta}$ is updated with a mini-batch (size equal J) of experiences $\{(\boldsymbol{s}_j, \mathrm{a}_j, \mathrm{r}_j, \boldsymbol{s}_{j+1})\}$ sampled from the replay memory.

Experience replay breaks the temporal correlation among experiences, and liberates RL agents from learning with transitions in the same order as they appear. In the ordinary experience replay, the J experiences in the mini-batch are uniformly sampled from the replay memory.

Prioritized Experience Replay

Different transitions have different contributions to the learning process, in terms of convergence. To further relieve the RL agent from learning with transitions in the same frequency as they appear and improve the learning efficiency, prioritized experience replay is introduced [12, 13].

Each transition j in the replay memory, i.e., $(\boldsymbol{s}_j, \mathrm{a}_j, \mathrm{r}_j, \boldsymbol{s}_{j+1})$, is assigned with a priority, p_j, which is equal to the magnitude of TD error δ_j plus a very small value ϵ_1, given by

$$p_j = \delta_j + \epsilon_1. \tag{4.66}$$

The magnitude of the TD error is a measure of how much the RL agent can learn from a transition. The larger is the TD error, the more aggressive is the update on weights $\boldsymbol{\theta}$ of the evaluation DQN. The small value ϵ_1 prevents transitions with zero TD errors from never being revisited.

Priority is also placed on the most recent transition. At the end of decision epoch k, the new transition $(\boldsymbol{s}_k, \mathrm{a}_k, \mathrm{r}_k, \boldsymbol{s}_{k+1})$ is available. The initial priority p_k for transition k is set as the maximum priority among all the existing transitions stored in the replay memory, given by

$$p_k = \max_{j<k} p_j. \tag{4.67}$$

The priority for transition k is given by (4.67) at only epoch $k + 1$. For $k' > k + 1$, the priority is given by (4.66) with $j = k$.

With the priority assigned to each transition, the sampling probability of transition j is

$$\Pr(j) = \frac{p_j^{o_1}}{\sum_{j=1}^{\mathrm{M}} p_j^{o_1}} \tag{4.68}$$

where $\mathtt{M}$ is the size of replay memory and o_1 determines the level of prioritization. For transitions with higher priority, the sampling probability is higher.

Prioritized Experience Replay with Penalty-Awareness

In the VNF migration problem, it is desired that the deep Q-learning algorithm converges to a solution without resource overloading penalty in the whole episode. However, such experiences are rare at the early learning stage with a lot of exploration, especially if an episode contains a large number of transitions. To learn more from such rare and desired experiences, we extend the prioritized experience replay technique to consider penalty-awareness. For the transitions in the replay memory except the most recent one, we place more priority on transitions with both high absolute TD error and zero penalty, with p_j slightly modified as

$$p_j = \omega^{(\mathrm{TD})} |\delta_j| + \left(1 - \omega^{(\mathrm{TD})}\right)\left(1 - \mathrm{f}_j^{(\mathrm{P})}\right) + \epsilon_1 \tag{4.69}$$

where $\omega^{(\mathrm{TD})} \in [0, 1]$ is a parameter controlling the relative importance of TD error and penalty avoidance, and $0 < \epsilon_1 \ll 1$ is a very small constant. In practice, we select $\omega^{(\mathrm{TD})}$ close to 1, e.g., 0.99, to incorporate penalty-awareness without significant degradation on convergence speed.

To enable the priority awareness, the penalty flag, $\mathsf{f}_j^{(\mathrm{P})}$ in (4.58), is recorded at learning step j, and a five-tuple transition $(\boldsymbol{s}_j, \mathrm{a}_j, \mathrm{r}_j, \mathsf{f}_j^{(\mathrm{P})}, \boldsymbol{s}_{j+1})$ instead of the original four-tuple transition, $(\boldsymbol{s}_j, \mathrm{a}_j, \mathrm{r}_j, \boldsymbol{s}_{j+1})$, is stored in the replay memory, for priority calculation based on (4.69).

The prioritization leads to a loss of diversity, which can be corrected with an importance-sampling weight w_j, given by [13]

$$\mathrm{w}_j = (B \cdot \Pr(j))^{-o_2} \,/\, \max_{j'}(\mathrm{w}_{j'}) \tag{4.70}$$

where o_2 controls the level of compensation. Correspondingly, the TD error δ_j is replaced by a weighted TD error $\mathrm{w}_j\delta_j$ in a gradient descent step with transition j, as given in Line 13 of Algorithm 4.1.

4.5 Performance Evaluation

4.5.1 Simulation System Setup

We consider a local network segment with one VNF, which can be placed at one of 6 candidate NFV nodes in $\{n_0, n_1, n_2, n_3, n_4, n_5\}$. At the beginning of each episode in RL, the initial placement of the VNF is at NFV node n_0. The processing capacities of all NFV nodes are the same, given by $\mathsf{R}_n = 125000$ packet/s. The background resource utilization factor of each NFV node varies between 10% and 90% in different patterns, as shown in Fig. 4.13.

For performance evaluation, we use real-world traffic traces measured at Internet backbone links by the MAWI working group of the WIDE project [14]. Packet-level information is provided in the traffic traces, including the packet arrival timestamps. We select two 48-h-long traffic traces collected from the transit link of WIDE backbone connecting the upstream ISP, which span the days of 2018/05/09, 2018/05/10, 2019/04/09, and 2019/04/10. We extract the packets of Hypertext Transfer Protocol (HTTP) application with port number 443 from the 4-day traffic traces, and constitute a 4-day long aggregate service flow.

We select the small interval length as $T_S = 100$ ms and the medium interval length as $T_M = 20$ s, and set the prior average run length $\bar{l} = \frac{3600}{T_M}$ in the BOCPD algorithm, corresponding to 1 h in time duration. The two thresholds, Υ_l and Υ_d, for deterministic change point detection based on the posterior run length distribution, are set as 10 and 5%, respectively. Under the simulation settings, 25, 20, 26, and 24 change points are detected in the four daily traffic traces, respectively, implying that resource demand prediction and VNF migration are triggered in an hour-level timescale.

For the *look-back* traffic parameter learning, we select $i_0 = 4$ to collect 800 small-timescale traffic samples before each detected change point, which gives high

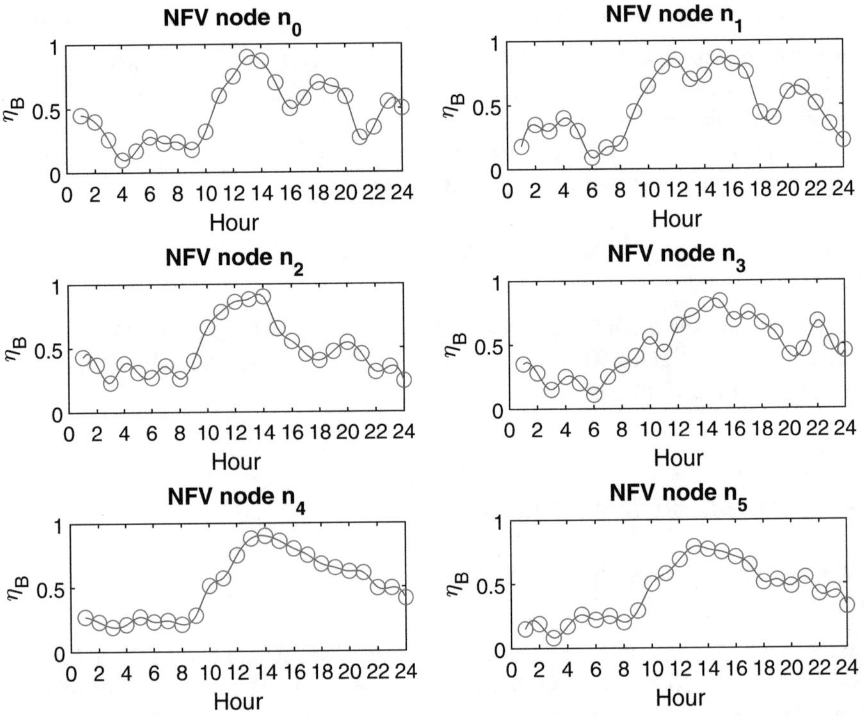

Fig. 4.13 Background resource utilization factors of candidate NFV nodes

computation efficiency while achieving a good accuracy. The accuracy of traffic parameter learning is evaluated with $t_0 = 1000$ small-timescale traffic samples.

For resource demand prediction, we consider two groups of probabilistic delay requirements with different delay bounds, i.e., $\mathsf{D} = 10$ ms and $\mathsf{D} = 50$ ms in $\Pr(\mathsf{d} > \mathsf{D}) \leq \varepsilon$. For each group, we consider three values for the maximum delay violation probability ε, including 0.1, 0.01, and 0.001.

For deep Q-learning, we use a DQN structure with one hidden layer of 20 neurons and Relu as the activation function, with important learning parameters summarized in Table 4.1. We set weighting factor $\omega^{(\mathrm{C})} = 0.6$ and penalty level $c^{(\mathrm{P})} = 5$ in the reward function (4.59). We consider 1 week for the time duration of an episode in RL, to have sufficient learning steps (decision epochs) with periodic dynamics in both change points and resource demands within an episode. With the available 4-day traffic traces, we artificially compose weekly traffic patterns in different episodes by combining the four daily traffic traces in random order. Each daily traffic trace corresponds to a group of change points and predicted resource demands over time. Hence, the change points and predicted resource demands for a weekly traffic trace are also combined in order, which are inputs (part of the state) of the RL algorithm. To increase diversity of the weekly traffic patterns, we add the change points and resource demands with noise in different randomness levels and

Table 4.1 List of parameters in deep Q-learning

Parameter	Definition	Value
α_Q	Learning rate	10^{-6}
γ	Discount factor	0.9
ϵ_2^{min}	Minimum exploration probability	0.01
$\epsilon_{2,\Delta}$	Step size of exploration probability	5×10^{-6}
$K_{\boldsymbol{\theta}}$	Number of steps to replace $\hat{\boldsymbol{\theta}}$ by $\boldsymbol{\theta}$	200
$\mathtt{M}$	Memory size	2000
J	Batch size	200
$\omega^{(\mathrm{TD})}$	Weight in the priority	0.99

Table 4.2 Traffic sets with different randomness levels

Traffic set	Change points	Resource demands
1	Detected	Predicted
2	Detected $\pm$ [0, 0.1] h	Predicted
3	Detected	Predicted $\pm$ [0%, 5%]
4	Detected $\pm$ [0, 0.1] h	Predicted $\pm$ [0%, 5%]

prepare four traffic sets as described in Table 4.2. The randomness level (in hour) around change points follows a uniform distribution in [0, 0.1], and the randomness level around resource demands follows a uniform distribution in [0%, 5%].

4.5.2 Simulation Results

We first check the Gaussianity of the real-world traffic in different timescales. Besides the small and medium timescales, we also consider a tiny time interval of length 1 ms. With the packet arrival timestamps, we can obtain the traffic time series in medium, small, and tiny timescales. For each traffic time series, we calculate the sample mean and standard deviation, based on which a centered normalized traffic time series (with mean equal to 0 and standard deviation equal to 1) can be obtained. Figure 4.14 presents the quantile-quantile (Q-Q) plots[4] for the distribution of centered normalized traffic time series in different timescales versus a standard Gaussian distribution. Here we use packets belonging to the same stationary traffic segment for Q-Q plotting. It shows that the traffic distributions in both small and medium timescales are approximately Gaussian with heavy tails, validating the assumption of Gaussianity in both the change point detection using medium-timescale traffic samples and the traffic parameter leaning using small-timescale

[4] A Q-Q plot is a probability plot, which compares two probability distributions by plotting their quantiles against each other. If the two probability distributions are similar, the points in the Q-Q plot will approximately lie on the diagonal line $y = x$.

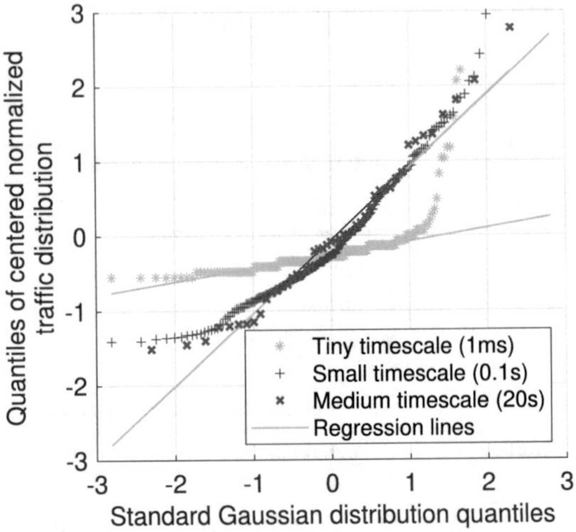

Fig. 4.14 Quantile-quantile (Q-Q) plots for different timescales

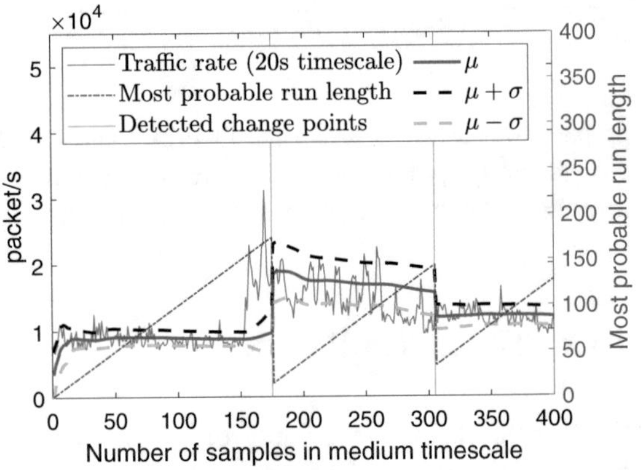

Fig. 4.15 Results of change point detection for a non-stationary traffic segment

traffic samples. However, the traffic distribution in a tiny timescale (1 ms) is more bursty and completely not Gaussian due to insufficient aggregation of packet arrivals in each tiny time interval.

Next, the simulation results for change point detection, traffic parameter learning, resource demand prediction, and VNF migration are presented.

4.5.2.1 Change Point Detection

Figure 4.15 shows results of change point detection for a non-stationary traffic segment in 8000 s, corresponding to 400 medium time intervals. A zigzag trend is observed for the most probable run length. The detected change points are indicated by gray vertical lines. Online estimation of mean and standard deviation in a student-

t distribution corresponding to the most probable run length at each time step (in medium timescale) is a byproduct of change point detection. It is observed that both statistics are stable between the detected change points. We see that both conditions in (4.35) and (4.36) are satisfied for the two detected change points, i.e., the most probable run length drops by more than Υ_l, and the change in mean plus standard deviation is sufficiently large. We also observe that the change points are detected after the occurrence of statistical changes, which verifies the effectiveness of the *look-back* traffic parameter learning.

4.5.2.2 Traffic Parameter Learning

To evaluate the accuracy of the traffic parameter learning method, we simulate six groups of fBm traffic traces in discrete time with different traffic parameters and resource demands, as given in Table 4.3. The simulated fBm traffic is generated following a wavelet-based algorithm [15]. The length of a time unit is not specified. The resource demand to satisfy the QoS requirement Pr($\mathsf{d} > 0.01$ time units) ≤ 0.01 is denoted by $\mathsf{R}_{0.01}$. For each group of fBm traffic, 200 sample paths are generated, with 1000 traffic samples in each sample path. The traffic parameters of each sample path are estimated by the first 800 traffic samples using both the learning method and a classical benchmark method. In the benchmark method, the mean and variance are estimated by the sample mean and variance, and the Hurst parameter is estimated separately using a wavelet-based approach [15]. In the learning method, the three parameters are learned together, reaching a compromise among them to maximize the log-marginal likelihood function given in (4.43). The results of both methods are given in Table 4.3. It is observed that the mean and Hurst parameter given by both methods are close to the simulated parameters, but the standard deviation is not as accurate. However, the level of underestimation given by the learning method is much lower than that given by the benchmark method. The accuracy of traffic parameter learning is also evaluated by the average prediction error, E, for the last 200 traffic samples in each sample path, as given in Table 4.3.

Table 4.3 Traffic parameters and resource demands of simulated fBm traffic

Group	Simulated				Estimated (benchmark)				Learned				
	λ	σ	H	$\mathsf{R}_{0.01}$	$\bar{\lambda}$	$\bar{\sigma}$	$\bar{H}$	$\mathsf{R}_{0.01}$	$\bar{\lambda}$	$\bar{\sigma}$	$\bar{H}$	$\mathsf{R}_{0.01}$	E
1	800	200	0.7	1429.6	797.9	216.3	0.698	1490.9	795.4	218.8	0.693	1501.4	0.1507
2	800	200	0.8	1403.4	804.4	191.4	0.796	1378.2	798.6	198.6	0.793	1397.8	0.1373
3	800	200	0.9	1422.5	801.3	182.4	0.894	1363.1	798.8	210.6	0.893	1453.2	0.1246
4	900	300	0.7	1895.7	898.5	325.6	0.697	1997.1	895.8	327.3	0.693	2008.3	0.1489
5	900	300	0.8	1836.4	896.9	287.5	0.796	1791.4	895.3	296.4	0.792	1822.0	0.1380
6	900	300	0.9	1848.7	899.6	271.8	0.895	1752.1	898.9	310.8	0.891	1879.0	0.1218

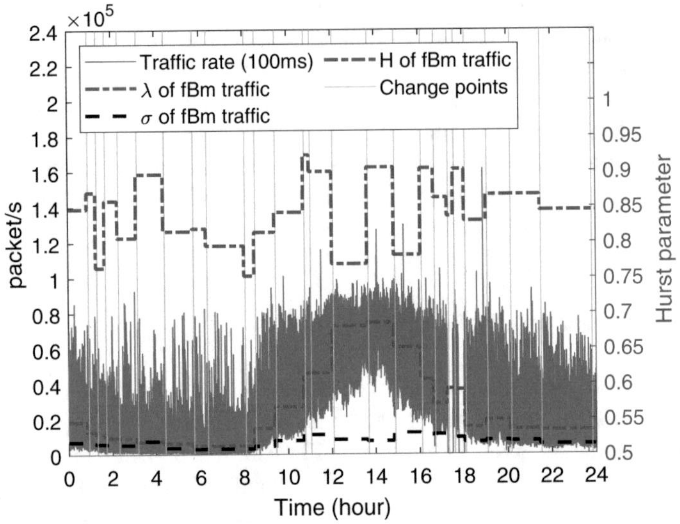

Fig. 4.16 Learned traffic parameters of daily traffic

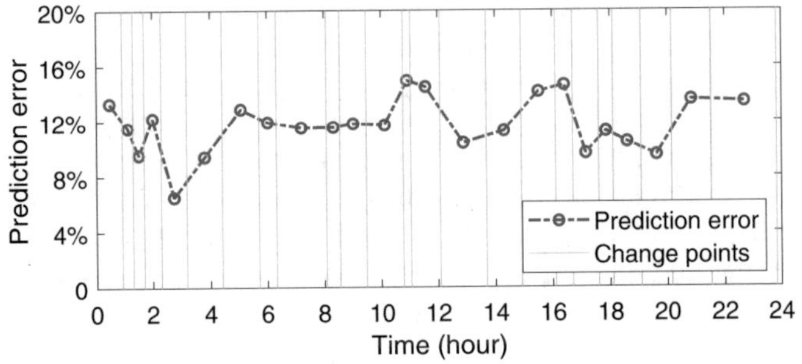

Fig. 4.17 Evaluation of traffic parameter learning accuracy for daily traffic

Figure 4.16 shows results of the change-point-driven traffic parameter learning scheme for a real-world daily traffic trace. The detected change points shown in gray vertical lines identify different stationary traffic segments. For each stationary traffic segment, the three learned fBm traffic parameters $\{\lambda\ (k)\ , \sigma\ (k)\ , H\ (k)\}$ are plotted. We observe that the Hurst parameter is within $[0.5, 1)$, indicating self-similarity and LRD of the traffic. The average traffic sample prediction error evaluated by $t_0 = 1000$ traffic samples in each identified stationary traffic segment is plotted in Fig. 4.17. It can be seen that the average prediction errors for the real-world traffic trace is comparable to that of the simulated fBm traffic traces given in Table 4.3.

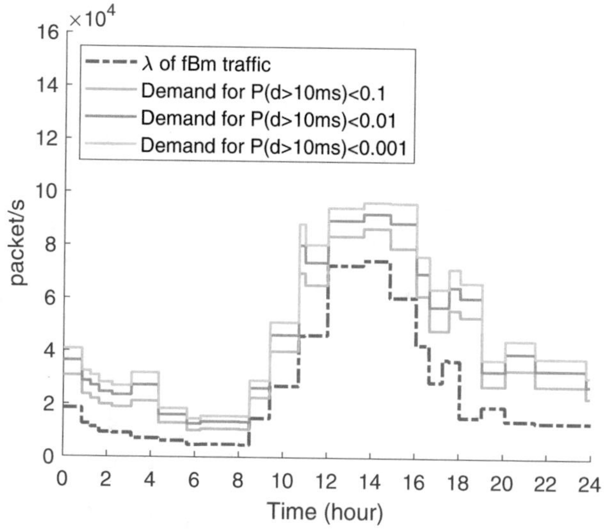

Fig. 4.18 Resource demand prediction results for daily traffic

4.5.2.3 Resource Demand Prediction

The predicted resource demands for the identified stationary traffic segments in the daily traffic trace are given in Fig. 4.18, for QoS requirements $\Pr(\mathsf{d} > 10\,\text{ms}) \leq \varepsilon$ with $\varepsilon = 0.1, 0.01$, and 0.001. As expected, the resource demand is greater than the average traffic rate and increases when ε decreases.

To evaluate QoS performance of the resource demand prediction scheme, we conduct packet-level simulations using the python Simpy package, to gather sufficient packet delay information for a smooth characterization of the VNF packet processing delay distribution, with a 60 s-long stationary traffic segment from the real-world traffic trace as the VNF traffic input. Different amount of resources are allocated to the VNF according to the predicted resource demands for different QoS requirements. For simplicity, we use R_ε to represent the predicted resource demand for a probabilistic delay guarantee, i.e., $\Pr(\mathsf{d} > \mathsf{D}) \leq \varepsilon$. Since traffic parameter learning is performed in 0.1 s timescale, traffic burstiness in time granularities smaller than 0.1 s cannot be captured. Hence, we use both the real packet arrival trace and a less-bursty synthesized packet arrival trace for QoS evaluation. In the synthesized packet arrival trace, the numbers of packet arrivals in 0.1 s timescale are the same as the real packet arrival trace, but the packet inter-arrival time within each 0.1 s time interval follows an exponential distribution. Figure 4.19 shows the distribution of VNF packet processing delay for both traffic traces. Two groups of delay requirements with different delay bounds, i.e., $\mathsf{D} = 10\,\text{ms}$ and $\mathsf{D} = 50\,\text{ms}$, are used for QoS evaluation. In each group, ε is set as 0.1, 0.01, and 0.001. For the same QoS requirement, the amount of resources allocated for both traffic are the same. However, the delay performance of the synthesized traffic is better than that of the

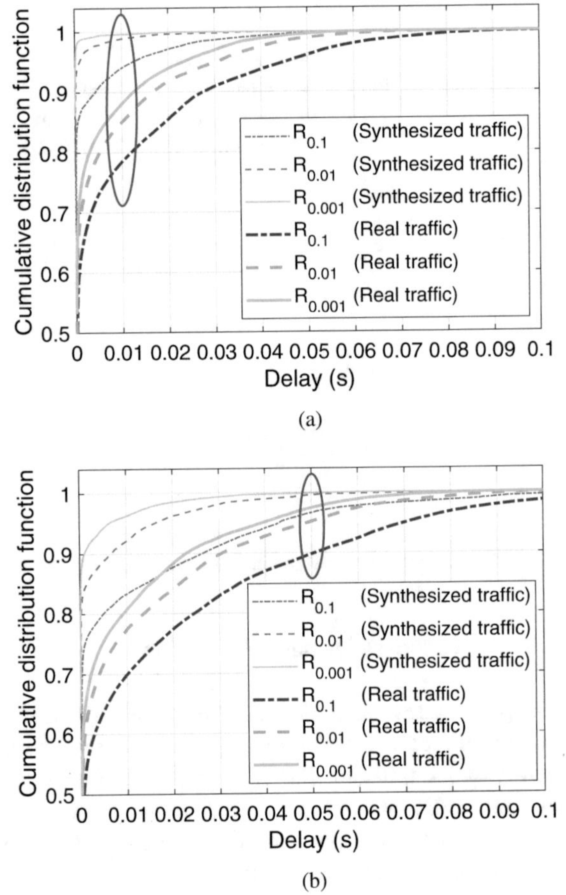

Fig. 4.19 Distribution of VNF packet processing delay for both the synthesized traffic and the real traffic. (**a**) $D = 10\,\text{ms}$. (**b**) $D = 50\,\text{ms}$

real traffic, due to less traffic burstiness in time granularities smaller than 0.1 s. For the synthesized traffic, the delay violation probability is within the corresponding upper limits. For the real traffic, the delay violation probability occasionally exceeds the required upper limit, especially for the stringent QoS requirements such as $\Pr(\mathsf{d} > 10\,\text{ms}) \leq 0.001$, due to traffic burstiness in time granularities below 0.1 s.

In addition, we compare the QoS performance between the fBm model based resource demand prediction scheme and a benchmark M/M/1 model based counterpart. Both methods use the learned traffic parameters for resource demand prediction. In the fBm based scheme, the resource demand is predicted from all three learned traffic parameters, i.e., $\{\lambda, \sigma, H\}$, based on the fBm resource provisioning model in Sect. 4.1.3. In the benchmark scheme, the resource demand is predicted from the first learned traffic parameter, i.e., λ, based on an M/M/1 resource provisioning model. For an M/M/1 queue with arrival rate λ (in packet/s) and service rate R (in packet/s), the delay violation probability is $\Pr(\mathsf{d} > \mathsf{D}) = e^{-(\mathsf{R}-\lambda)\mathsf{D}}$ [16]. Hence, the minimum amount of resources (in packet/s) to guarantee the QoS

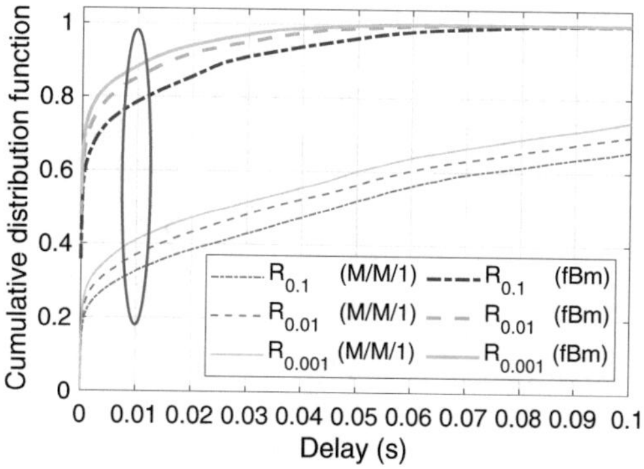

Fig. 4.20 QoS performance comparison between the fBm model and M/M/1 model based resource demand prediction schemes

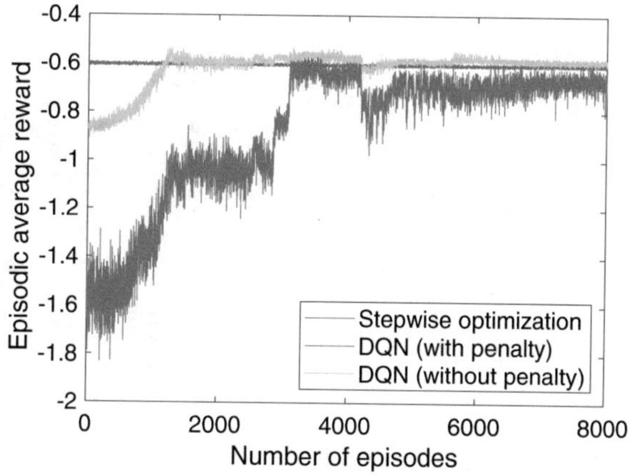

Fig. 4.21 Episodic average reward versus the episode number for D*Q*N

requirement $\Pr(d > D) \leq \varepsilon$ is $R_{S,min} = \lambda - \frac{\log \varepsilon}{D}$. Figure 4.20 shows the VNF packet delay distribution with the real packet arrivals and with different amount of resources allocated to the VNF, based on predicted resource demands given by the two models. A gap is observed between the delay performance of the two models. The fBm resource provisioning model, with the ability to capture the bursty nature of traffic, gives a better estimation of resource demands.

4.5.2.4 VNF Migration

The performance of the deep *Q*-learning algorithm with penalty-aware prioritized experience replay (PP-D*Q*N) is compared with two benchmark algorithms, i.e., deep *Q* learning with uniformly sampled experience replay (D*Q*N), and deep *Q*-learning with prioritized experience relay (P-D*Q*N). All three deep *Q*-learning algorithms are compared with a common benchmark, i.e., stepwise optimization, which solves a one-shot VNF migration problem at each decision epoch. The comparison is performed using traffic set 1 in Table 4.2.

Fig. 4.22 Episodic average reward versus the episode number for (**a**) P-DQN and (**b**) PP-DQN

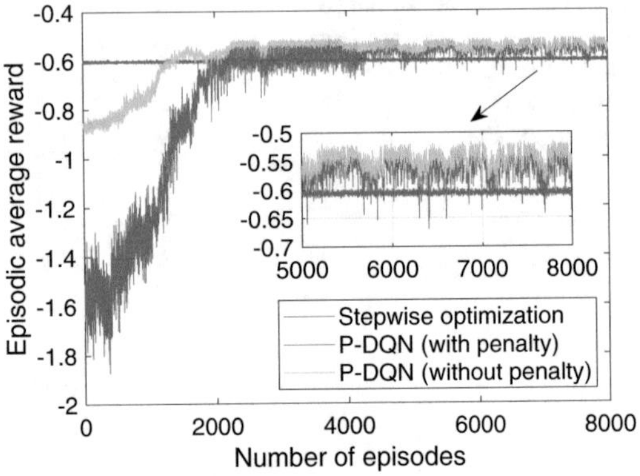

(a)

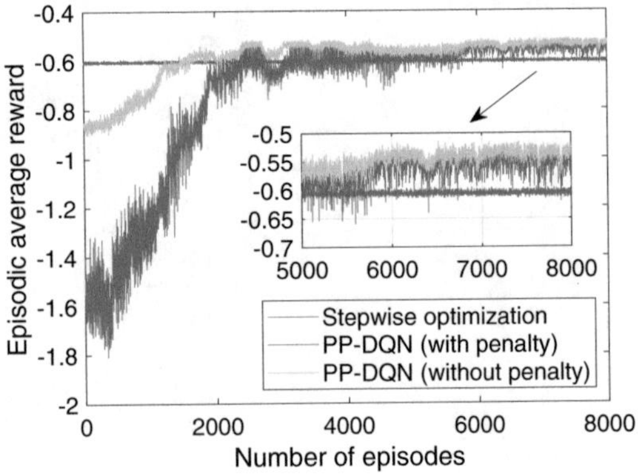

(b)

Figures 4.21 and 4.22 show the evolution of episodic average reward with respect to the number of episodes during the learning process, using the three deep Q-learning algorithms. Both the full reward including penalty and the partial reward without penalty are plotted, with a gap indicating the penalty. It is observed from Fig. 4.21 that DQN converges to a poor solution which is worse than the stepwise optimization benchmark in terms of episodic average reward. The penalty is high, inferring that the DQN does not learn a solution to minimize the resource overloading penalty in the long run. As shown in Fig. 4.22, both the P-DQN and PP-DQN algorithms take advantages of the prioritized experience replay for convergence to solutions that outperform the stepwise optimization benchmark in most time after convergence. It demonstrates that both P-DQN and PP-DQN after convergence can capture the daily and weekly traffic patterns (in both change points and resource demands) and background resource loading patterns at the candidate NFV nodes, and make intelligent VNF migration decisions accordingly. In contrast, when a VNF migration is required, the stepwise optimization benchmark favors VNF migration to a lightly loaded NFV node in the current decision epoch, which

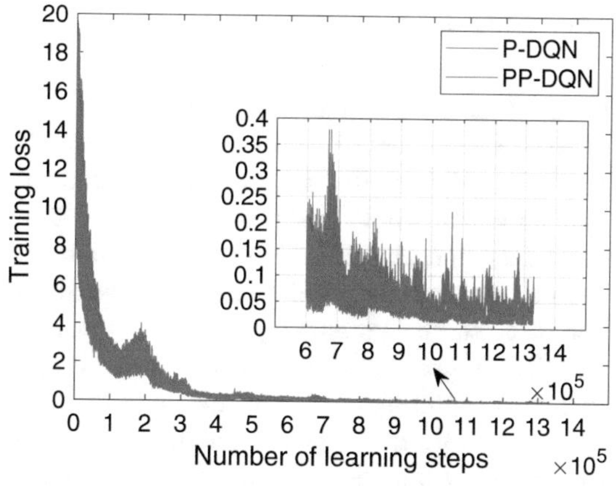

Fig. 4.23 Training loss of the evaluation Q networks

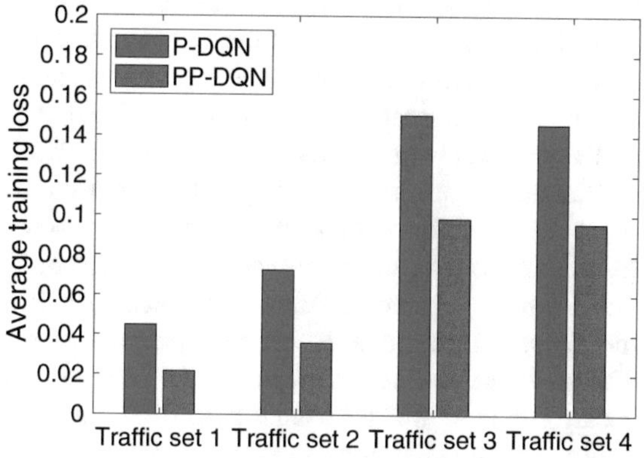

Fig. 4.24 Average training loss after convergence

can be heavily loaded in the following decision epochs. The PP-DQN achieves slightly more gain in terms of penalty suppression compared with P-DQN. The episodic average rewards (with penalty) of P-DQN and PP-DQN after convergence are -0.5502 and -0.5408 respectively.

Figure 4.23 shows that the training loss of PP-DQN as defined in (4.62) converges faster to a smaller value. We also examined the learning curve of PP-DQN with $\omega^{(\mathrm{TD})} = 0.5$, which gives an average training loss of 0.0155 after convergence, demonstrating the benefit of reducing weight $\omega^{(\mathrm{TD})}$ on further loss reduction. However, the benefit on additional reward improvement is not significant. To evaluate generalization of the penalty-aware deep Q-learning algorithm to similar traffic sets with different randomness, we compare the average training loss of both P-DQN and PP-DQN after convergence in Fig. 4.24, using four traffic sets with different randomness levels in change points and resource demands, as in Table 4.2. With more randomness especially in resource demand, the average training loss of both P-DQN and PP-DQN increases. However, the PP-DQN outperforms P-DQN for all the traffic sets.

4.6 Summary

In this chapter, we study a dynamic VNF scaling problem in a local network segment for probabilistic delay guarantee. One VNF is allowed to migrate among a set of candidate NFV nodes, with elastic resource scaling. We consider a non-stationary real-world traffic trace as the traffic input for the VNF. A change point detection algorithm is presented, which identifies the boundaries between consecutive stationary traffic segments based on the changes in traffic statistics. The detected change points provide triggers for both resource demand prediction and VNF migration decision. Under the assumption of fBm traffic model for each stationary traffic segment, a *look-back* traffic parameter learning scheme is used to infer three fBm traffic parameters including mean, standard deviation, and Hurst parameter, by training a Gaussian process regression model with fBm covariance function. The learned traffic parameters are then used as input for a resource provisioning model, to predict the resource demand for a given probabilistic delay requirement. With the detected change points and predicted resource demands, the dynamic VNF migration problem is formulated as an MDP, which is then solved by an RL approach with a penalty-aware deep Q-learning algorithm.

Simulation results demonstrate the effectiveness and accuracy of the change-point-driven traffic parameter learning and resource demand prediction method. Benefiting from a compromise among the three traffic parameters, the traffic parameter learning scheme achieves better accuracy than a benchmark method where the Hurst parameter is estimated separately from mean and standard deviation. The predicted resource demand by the fBm resource provisioning model is more accurate than an M/M/1 counterpart, from the perspective of QoS guarantee. Benefiting from a penalty-aware prioritized experience replay, the penalty-aware deep Q-learning algorithm achieves faster training loss reduction and higher cumulative reward which integrates the trade-off among load balancing, migration cost, and resource overloading penalty [17].

References

1. Krithikaivasan, B., Zeng, Y., Deka, K., Medhi, D.: ARCH-based traffic forecasting and dynamic bandwidth provisioning for periodically measured nonstationary traffic. IEEE/ACM Trans. Netw. **15**(3), 683–696 (2007)
2. Fraleigh, C., Tobagi, F., Diot, C.: Provisioning IP backbone networks to support latency sensitive traffic. In: Proc. IEEE INFOCOM, pp. 1871–1879 (2003)
3. Kim, J., Hwang, G.: Adaptive bandwidth allocation based on sample path prediction with Gaussian process regression. IEEE Trans. Wirel. Commun. **18**(10), 4983–4996 (2019)
4. Cheng, Y., Zhuang, W., Wang, L.: Calculation of loss probability in a finite size partitioned buffer for quantitative assured service. IEEE Trans. Commun. **55**(9), 1757–1771 (2007)
5. Kim, H.S., Shroff, N.B.: Loss probability calculations and asymptotic analysis for finite buffer multiplexers. IEEE/ACM Trans. Netw. **9**(6), 755–768 (2001)

6. Adams, R.P., MacKay, D.J.: Bayesian online changepoint detection. Tech. rep., University of Cambridge, Cambridge, UK (2007)
7. Murphy, K.P.: Conjugate Bayesian analysis of the Gaussian distribution. def **1**($2\sigma 2$), 16 (2007)
8. Bishop, C.M.: Pattern Recognition and Machine Learning. Springer, New York (2006)
9. Bayati, A., Asghari, V., Nguyen, K., Cheriet, M.: Gaussian process regression based traffic modeling and prediction in high-speed networks. In: Proc. IEEE GLOBECOM, pp. 1–7 (2016)
10. Williams, C.K., Rasmussen, C.E.: Gaussian Processes for Machine Learning, vol. 2. MIT Press, Cambridge (2006)
11. Mnih, V., Kavukcuoglu, K., Silver, D., Rusu, A.A., Veness, J., Bellemare, M.G., Graves, A., Riedmiller, M., Fidjeland, A.K., et al.: Human-level control through deep reinforcement learning. Nature **518**(7540), 529–533 (2015)
12. Xu, Z., Tang, J., Meng, J., Zhang, W., Wang, Y., Liu, C.H., Yang, D.: Experience-driven networking: A deep reinforcement learning based approach. In: Proc. IEEE INFOCOM, pp. 1871–1879 (2018)
13. Schaul, T., Quan, J., Antonoglou, I., Silver, D.: Prioritized experience replay. In: Proc. ICLR'16, pp. 1–7 (2016)
14. MAWI Working Group Traffic Archive (2021). http://mawi.wide.ad.jp/mawi/. Accessed 14 July 2021
15. Abry, P., Sellan, F.: The wavelet-based synthesis for fractional Brownian motion proposed by F. Sellan and Y. Meyer: remarks and fast implementation. Appl. Comput. Harmon. Anal. **3**(4), 377–383 (1996)
16. Kobayashi, H., Mark, B.L.: System Modeling and Analysis: Foundations of System Performance Evaluation. Pearson Education India (2009)
17. Qu, K., Zhuang, W., Ye, Q., Shen, X., Li, X., Rao, J.: Dynamic resource scaling for VNF over nonstationary traffic: a learning approach. IEEE Trans. Cogn. Commun. Netw. **7**(2), 648–662 (2021)

Chapter 5
Dynamic VNF Scheduling for Network Utility Maximization

5.1 System Model

5.1.1 Services

Consider multiple services in a time-slotted system where time is partitioned into equal-length time slots indexed by τ. The time slot length is T in second, which is given by the time quantum of the operating system (OS). Each service is in the form of VNF chain, originating from an ingress edge switch and traversing through a sequence of VNFs towards an egress edge switch. Let $\mathcal{R}$ denote the set of services. Let H_r be the number of VNFs in service $r \in \mathcal{R}$, and let $\mathcal{H}_r = \{1, \cdots, H_r\}$ be a set containing the index of VNFs of service $r \in \mathcal{R}$. Denote the h-th ($h \in \mathcal{H}_r$) VNF in SFC $r \in \mathcal{R}$ as $V_h^{(r)}$. Under the assumption of same virtualization platform at the NFV nodes, let $P_h^{(r)}$ denote the processing density of VNF $V_h^{(r)}$ at any NFV nodes in cycle/packet, which is the CPU resource demand (in cycle/s) of VNF $V_h^{(r)}$ for one packet/s of processing rate [1, 2]. We assume that the exogenous packet arrivals for each service occur only at the ingress edge switch. The arrival process is stationary and ergodic with mean rate $\bar{a}_r$ (in packet per time slot) for service $r \in \mathcal{R}$. Let $A^{(r)}(\tau)$ denote the number of packets that arrive at the τ-th time slot for service $r \in \mathcal{R}$, which is highly dynamic and unpredictable. Assume that $A^{(r)}(\tau)$ is upper bounded by a finite maximum value $A_{max}^{(r)}$ for service $r \in \mathcal{R}$. The QoS requirement of service $r \in \mathcal{R}$ is represented by two parameters, M_r and ε_r, where M_r is the E2E deadline (in time slots) for each packet of service $r \in \mathcal{R}$, and ε_r specifies the maximum packet dropping ratio of service $r \in \mathcal{R}$ due to E2E deadline violation. For a deadline-constrained service, a packet becomes useless and should be dropped once the packet delay violates the E2E deadline constraint. For service $r \in \mathcal{R}$, let $\bar{f}_r$ denote the throughput in packet per time slot, which is the average number of packets successfully delivered within E2E deadline M_r to the egress edge switch in a time slot. To satisfy the QoS requirement, it requires that $\bar{f}_r \geq \bar{a}_r(1 - \varepsilon_r)$.

W. Zhuang, K. Qu, *Dynamic Resource Management in Service-Oriented Core Networks*, Wireless Networks, https://doi.org/10.1007/978-3-030-87136-9_5

5.1.2 Network Model

We consider a core network with a set $\mathcal{N}$ of NFV nodes interconnected by virtual links. The CPU processing resource budget at NFV node $n \in \mathcal{N}$ is C_n in cycle per time slot. Assume that there are sufficient transmission resources in the network for virtual link provisioning to enable communications among the NFV nodes. The services in set $\mathcal{R}$ are embedded in the network, with VNF placement at NFV nodes and traffic routing over virtual links. Here, we consider fixed VNF placement and traffic routing. Let x_n^{rh} be a binary parameter, with $\mathrm{x}_n^{rh} = 1$ if VNF $V_h^{(r)}$ is placed at NFV node $n \in \mathcal{N}$, and $\mathrm{x}_n^{rh} = 0$ otherwise. Let $\mathcal{V}_n = \{(r, h)|r \in \mathcal{R}, h \in \mathcal{H}_r, \mathrm{x}_n^{rh} = 1\}$ be a set containing the index of all VNFs placed at NFV node $n \in \mathcal{N}$, with $(r, h) \in \mathcal{V}_n$ denoting VNF $V_h^{(r)}$.

Each NFV node has multiple CPU cores in general. Different VNFs can be scheduled in parallel at different CPU cores, but can only be scheduled sequentially for each CPU core. For simplicity, we consider only NFV nodes with a single CPU. Under this assumption, at most one VNF in set $\mathcal{V}_n$ can be scheduled for packet processing at NFV node $n \in \mathcal{N}$ during each time slot. Let $z_h^{(r)}(\tau)$ be a binary VNF scheduling decision variable, with $z_h^{(r)}(\tau) = 1$ if VNF $V_h^{(r)}$ is scheduled for packet processing at the corresponding NFV node $n \in \mathcal{N}$ with $x_n^{rh} = 1$ during time slot τ, and $z_h^{(r)}(\tau) = 0$ otherwise. Under the single-CPU assumption, we have at most one VNF scheduled during time slot τ at NFV node $n \in \mathcal{N}$, given by

$$\sum_{(r,h)\in\mathcal{V}} x_n^{rh} z_h^{(r)}(\tau) = \sum_{(r,h)\in\mathcal{V}_n} z_h^{(r)}(\tau) \leq 1, \quad \forall n \in \mathcal{N}. \tag{5.1}$$

5.1.3 Discrete-Time VNF Packet Processing Queueing Model

5.1.3.1 Physical Packet Processing Queues

At each NFV node $n \in \mathcal{N}$, a separate physical packet processing queue is maintained for each VNF in set $\mathcal{V}_n$. Let $q_h^{(r)}(\tau)$ denote the number of packets in the physical processing queue associated with VNF $V_h^{(r)}$ at the beginning of time slot τ. Let $S_h^{(r)}(\tau)$ be the number of packets processed from the queue of VNF $V_h^{(r)}$ during time slot τ, and let $D_h^{(r)}(\tau)$ denote the number of packets dropped from the queue of VNF $V_h^{(r)}$ during time slot τ due to packet deadline violation.

Under the assumption that the virtual link delay is negligible, all packets processed by VNF $V_h^{(r)}$ during time slot τ can arrive at the downstream VNF $V_{h+1}^{(r)}$ for $h \in \mathcal{H}_r \backslash \{H_r\}$ or at the egress edge switch for $h = H_r$ before the beginning of time slot $\tau + 1$. Existing studies usually assume that a packet can be processed by at most one VNF in a chain during one time slot [2]. Under both assumptions,

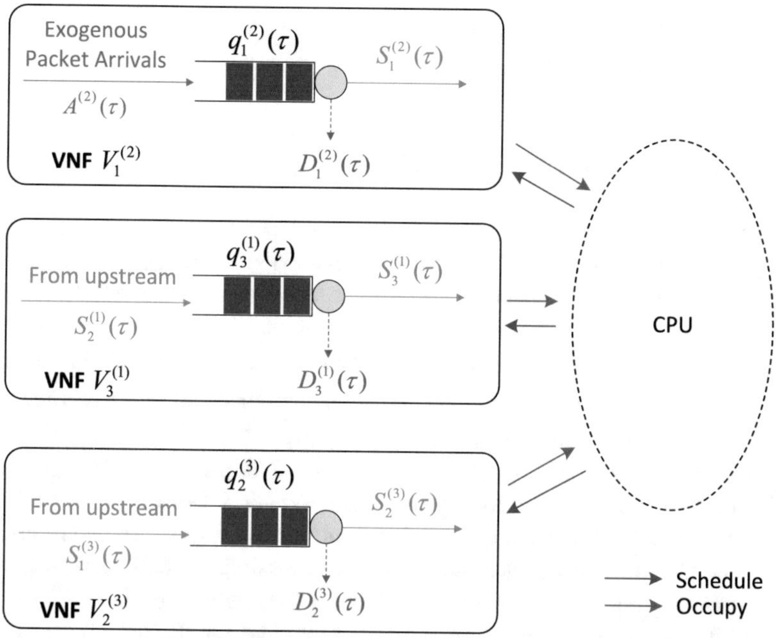

Fig. 5.1 An illustration of physical packet processing queues for VNFs at an NFV node

for $h \in \mathcal{H}_r \backslash \{H_r\}$, the packets processed from VNF $V_h^{(r)}$ during time slot τ are the new packet arrivals for the downstream VNF $V_{h+1}^{(r)}$ at the beginning of time slot $\tau+1$, which cannot be processed at the downstream VNF $V_{h+1}^{(r)}$ until time slot $\tau+1$. Correspondingly, the queue length of VNF $V_h^{(r)}$ at the beginning of time slot $\tau+1$, i.e., $q_h^{(r)}(\tau+1)$, is updated as

$$q_h^{(r)}(\tau+1) = \left[q_h^{(r)}(\tau) - S_h^{(r)}(\tau) - D_h^{(r)}(\tau)\right]^+ + S_{h-1}^{(r)}(\tau)\mathbb{1}_{\{h>1\}}$$
$$+A^{(r)}(\tau)\mathbb{1}_{\{h=1\}}, \quad \forall r \in \mathcal{R}, \quad \forall h \in \mathcal{H}_r \tag{5.2}$$

where $\mathbb{1}$ is the indicator function, equal to 1 only if the condition inside the bracket is true and 0 otherwise. For VNF $V_h^{(r)}$, the packet departures during time slot τ are composed of both the processed and dropped packets from the queue of VNF $V_h^{(r)}$ during the time slot. If $h = 1$, VNF $V_h^{(r)}$ is the first VNF in the chain, and the packet arrivals are the exogenous packet arrivals for the service. If $h > 1$, the packet arrivals of VNF $V_h^{(r)}$ at the beginning of time slot $\tau+1$ are the processed packets from the upstream VNF $V_{h-1}^{(r)}$ during the previous time slot τ.

Figure 5.1 illustrates three physical packet processing queues associated with VNFs $V_1^{(2)}$, $V_3^{(1)}$ and $V_2^{(3)}$ at an NFV node with a single CPU. During time slot τ, at most one physical queue can be scheduled, with $S_h^{(r)}(\tau) > 0$. The non-scheduled physical queues all have $S_h^{(r)}(\tau) = 0$.

5.1.3.2 Delay-Aware Virtual Packet Processing Queues

For service $r \in \mathcal{R}$, a new packet admitted to the first VNF $V_1^{(r)}$ in the chain has a zero packet delay. The delay (in number of time slots) of a packet is increased by 1, for each time slot. If a packet of service $r \in \mathcal{R}$ cannot be successfully delivered to the egress edge switch before the E2E deadline of M_r, it expires and should be dropped. Since there are $h - 1$ upstream VNFs preceding VNF $V_h^{(r)}$ in the chain, the delay of any packets at VNF $V_h^{(r)}$ is larger than or equal to $h - 1$ time slots, i.e., the residual packet lifetime does not exceed $M_r - h + 1$ time slots at VNF $V_h^{(r)}$ before deadline violation. Also, since there are $H_r - h$ downstream VNFs after VNF $V_h^{(r)}$, a packet with a residual lifetime less than $H_r - h + 1$ time slots at VNF $V_h^{(r)}$ cannot successfully reach the egress edge switch before expiry. To avoid resource inefficiency for processing such packets at the downstream VNFs, a packet with a residual lifetime equal $H_r - h + 1$ time slots during a certain time slot at VNF $V_h^{(r)}$ is dropped if it is not processed before the end of the time slot. Hence, all the existing packets at VNF $V_h^{(r)}$ during any time slot have a residual lifetime $m \in \mathcal{M}_h^{(r)}$, where $\mathcal{M}_h^{(r)}$ is a set given by

$$\mathcal{M}_h^{(r)} = \{H_r - h + 1, \cdots, M_r - h + 1\}, \quad \forall r \in \mathcal{R},\ h \in \mathcal{H}_r. \tag{5.3}$$

For example, we have $\mathcal{M}_1^{(r)} = \{H_r, \cdots, M_r\}$ for the first VNF $V_1^{(r)}$ of service $r \in \mathcal{R}$. At VNF $V_h^{(r)}$, all the packets with residual lifetime $m = H_r - h + 1$ are referred to as urgent packets, and other packets with residual lifetime $m \in \mathcal{M}_h^{(r)} \backslash \{H_r - h + 1\}$ are non-urgent packets.

According to the packet residual lifetime $m \in \mathcal{M}_h^{(r)}$, the physical packet processing queue associated with VNF $V_h^{(r)}$, $q_h^{(r)}(\tau)$, can be decomposed into $|\mathcal{M}_h^{(r)}|$ delay-aware virtual packet processing queues, denoted by $\{Q_{h,m}^{(r)}(\tau), \forall m \in \mathcal{M}_h^{(r)}\}$, with $q_h^{(r)}(\tau) = \sum_{m \in \mathcal{M}_h^{(r)}} Q_{h,m}^{(r)}(\tau)$. Each delay-aware virtual packet processing queue corresponds to a residual packet lifetime of $m \in \mathcal{M}_h^{(r)}$, with $Q_{h,m}^{(r)}(\tau)$ denoting the number of packets with residual lifetime $m \in \mathcal{M}_h^{(r)}$ at VNF $V_h^{(r)}$ at the beginning of time slot τ. Let $S_{h,m}^{(r)}(\tau)$ be the number of packets with residual lifetime $m \in \mathcal{M}_h^{(r)}$ that are processed at VNF $V_h^{(r)}$ during time slot τ, with $S_h^{(r)}(\tau) = \sum_{m \in \mathcal{M}_h^{(r)}} S_{h,m}^{(r)}(\tau)$. Let $D_h^{(r)}(\tau)$ denote the number of packets dropped from VNF $V_h^{(r)}$ during time slot τ. Then, the physical queue decomposition for both the first VNF with $h = 1$ and the remaining VNFs with $h > 1$ of service $r \in \mathcal{R}$ is illustrated in Figures 5.2 and 5.3 respectively.

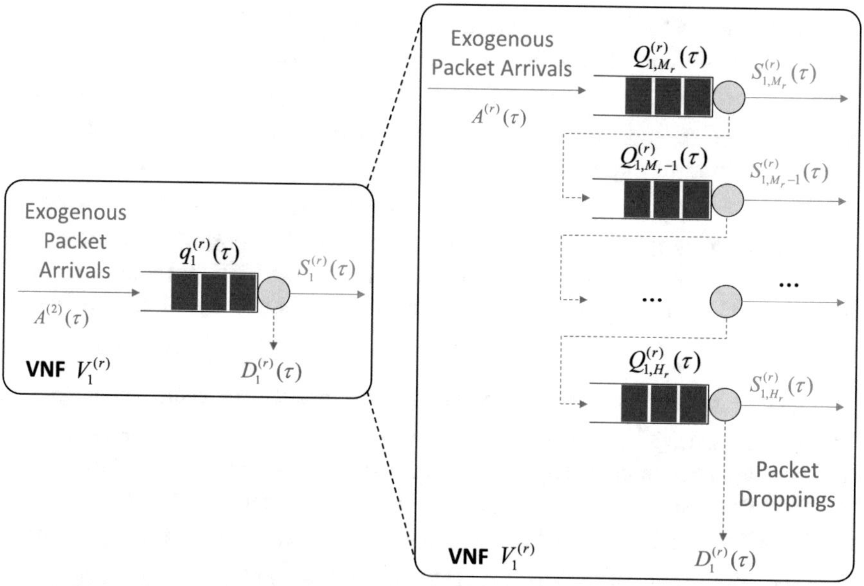

Fig. 5.2 An illustration of delay-aware virtual packet processing queues for VNF $V_1^{(r)}$

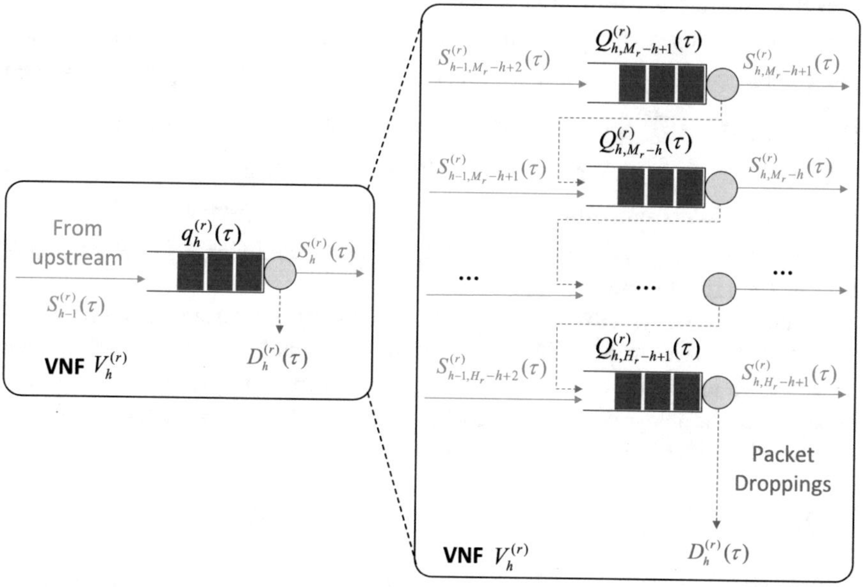

Fig. 5.3 An illustration of delay-aware virtual packet processing queues for VNF $V_h^{(r)}$ $(h > 1)$

For the first VNF $V_1^{(r)}$ of service $r \in \mathcal{R}$, the newly arrived packets with a full residual lifetime of M_r are all placed into virtual packet processing queue associated with $m = M_r$. During time slot τ, $S_{1,M_r}^{(r)}(\tau)$ packets with residual lifetime $m = M_r$ are processed and transmitted to the second VNF $V_2^{(r)}$. The residual lifetime of these packets is decreased by 1 to $M_r - 1$ at the beginning of time slot $\tau + 1$. Hence, it is equivalent to transfer $S_{1,M_r}^{(r)}(\tau)$ packets to the virtual packet processing queue $Q_{2,M_r-1}^{(r)}(\tau)$ at the second VNF, during time slot τ. The unprocessed packets with $m = M_r$ remain at the first VNF, but with a residual lifetime of $M_r - 1$ at the beginning of the next time slot $\tau + 1$, which is equivalent to transfer $\left[Q_{1,M_r}^{(r)}(\tau) - S_{1,M_r}^{(r)}(\tau)\right]^+$ packets to the virtual packet processing queue $Q_{1,M_r-1}^{(r)}(\tau)$ during time slot τ. Similarly, the packets in each virtual packet processing queue $Q_{1,m}^{(r)}(\tau)$ with $H_r - 1 \le m \le M_r - 1$ are split into processed and unprocessed parts, indicated by the blue and red arrows respectively in Fig. 5.2. The bottom virtual packet processing queue with $m = H_r$ contains the urgent packets at VNF $V_1^{(r)}$. The unprocessed part of the urgent packets should be dropped due to foreseen deadline violation, represented by $D_1^{(r)}(\tau) = \left[Q_{1,H_r}^{(r)}(\tau) - S_{1,H_r}^{(r)}(\tau)\right]^+$ for VNF $V_1^{(r)}$.

For the remaining VNFs with $h > 1$ illustrated in Fig. 5.3, except the top virtual packet processing queue with $m = M_r - h + 1$, there are two sources of packet arrivals for each virtual packet processing queue at VNF $V_h^{(r)}$. For virtual packet processing queue $Q_{h,m}^{(r)}(\tau)$ $(H_r - h + 1 \le m < M_r - h + 1)$, the first source is $\left[Q_{h,m+1}^{(r)}(\tau) - S_{h,m+1}^{(r)}(\tau)\right]^+$ unprocessed packets from the upper neighboring virtual queue $Q_{h,m+1}^{(r)}(\tau)$ at the same VNF, and the second source is $S_{h-1,m+1}^{(r)}(\tau)$ processed packets from virtual queue $Q_{h-1,m+1}^{(r)}(\tau)$ at the upstream VNF $V_{h-1}^{(r)}$. The top virtual packet processing queue $Q_{h,M_r-h+1}^{(r)}(\tau)$ has a single source of packet arrivals from the upstream VNF.

Figure 5.4 illustrates the delay-aware virtual packet processing queues from end to end for a service with 3 VNFs and packet E2E deadline of 6 time slots. Each VNF is associated with $|\mathcal{M}_h^{(r)}| = 4$ virtual packet processing queues. The queueing dynamics of the delay-aware virtual packet processing queues of service $r \in \mathcal{R}$ from end to end are updated as

$$
\begin{aligned}
&Q_{1,M_r}^{(r)}(\tau+1) = A^{(r)}(\tau), && \forall r \in \mathcal{R} \\
&Q_{1,m}^{(r)}(\tau+1) = \left[Q_{1,m+1}^{(r)}(\tau) - S_{1,m+1}^{(r)}(\tau)\right]^+, && \forall r \in \mathcal{R}, \quad \forall m \in \mathcal{M}_1^{(r)} \backslash \{M_r\} \\
&Q_{h,M_r-h+1}^{(r)}(\tau+1) = S_{h-1,M_r-h+2}^{(r)}(\tau), && \forall r \in \mathcal{R}, \quad \forall h \in \mathcal{H}_r \backslash \{1\} \\
&Q_{h,m}^{(r)}(\tau+1) = \left[Q_{h,m+1}^{(r)}(\tau) - S_{h,m+1}^{(r)}(\tau)\right]^+
\end{aligned}
$$

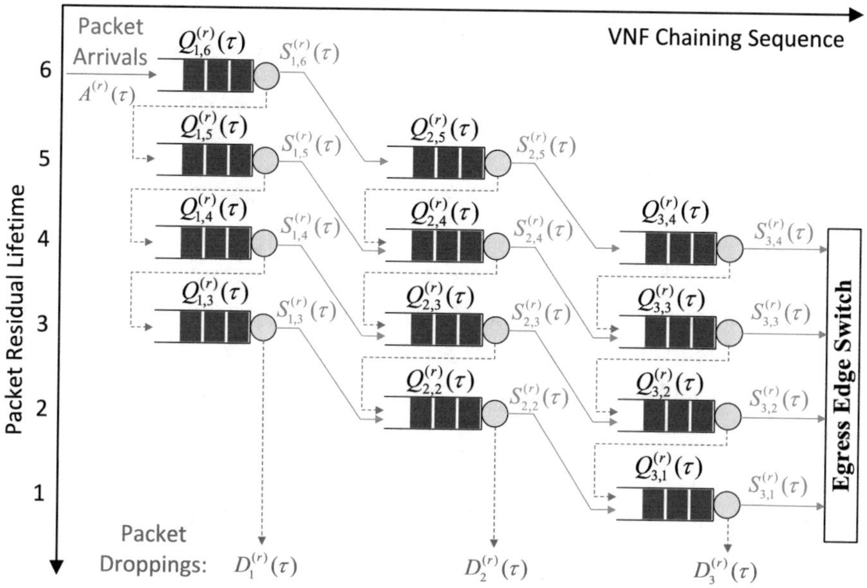

Fig. 5.4 An illustration of delay-aware virtual packet processing queueing model for service r with $H_r = 3$ and $M_r = 6$

$$+ S_{h-1,m+1}^{(r)}(\tau), \qquad \forall r \in \mathcal{R}, \quad \forall h \in \mathcal{H}_r \backslash \{1\},$$
$$\forall m \in \mathcal{M}_h^{(r)} \backslash \{M_r - h + 1\} \tag{5.4}$$

which combines the physical packet processing queue dynamics in (5.2) and the packet delay evolution over time. The packets with foreseen deadline violation are dropped at each VNF, with $D_h^{(r)}(\tau)$ equal to the number of unprocessed urgent packets at VNF $V_h^{(r)}$ during time slot τ, represented by

$$D_h^{(r)}(\tau) = \left[Q_{h,H_r-h+1}^{(r)}(\tau) - S_{h,H_r-h+1}^{(r)}(\tau)\right]^+, \quad \forall r \in \mathcal{R}, \quad \forall h \in \mathcal{H}_r. \tag{5.5}$$

The average E2E delay (in second) of the timely delivered packets of service $r \in \mathcal{R}$ to the egress edge switch, denoted by $\bar{d}_r$, is calculated as

$$\bar{d}_r = \sum_{m \in \mathcal{M}_{H_r}^{(r)}} \left[(M_r - m + 1)\, T \frac{\lim_{\Gamma\to\infty} \frac{1}{\Gamma} \sum_{\tau=0}^{\Gamma-1} \mathbb{E}\left\{S_{H_r,m}^{(r)}(\tau)\right\}}{\sum_{m \in \mathcal{M}_{H_r}^{(r)}} \lim_{\Gamma\to\infty} \frac{1}{\Gamma} \sum_{\tau=0}^{\Gamma-1} \mathbb{E}\left\{S_{H_r,m}^{(r)}(\tau)\right\}} \right],$$
$$\forall r \in \mathcal{R} \tag{5.6}$$

where T is the time slot length in second, Γ is the total number of time slots, and $\mathbb{E}$ denotes expectation over the randomness in packet arrivals, VNF scheduling, and packet processing.

5.1.4 Per-VNF FCFS Prioritized Packet Processing

If $z_h^{(r)}(\tau) = 1$, VNF $V_h^{(r)}$ is scheduled for packet processing during time slot τ, during which the CPU of NFV node $n \in \mathcal{N}$ with $x_n^{rh} = 1$ serves VNF $V_h^{(r)}$ in its maximum processing rate C_n (in cycle per time slot). With processing density $P_h^{(r)}$, the number of packets processed at VNF $V_h^{(r)}$ during time slot τ, i.e., $S_h^{(r)}(\tau)$, is given by

$$S_h^{(r)}(\tau) = z_h^{(r)}(\tau) \cdot \min\left(Q_h^{(r)}(\tau), \left\lfloor \frac{\sum_{n\in\mathcal{N}} x_n^{rh} C_n}{P_h^{(r)}} \right\rfloor\right). \tag{5.7}$$

If $z_h^{(r)}(\tau) = 0$, we have $S_h^{(r)}(\tau) = 0$. Otherwise, $S_h^{(r)}(\tau)$ is determined by the smaller value between the corresponding physical packet processing queue length $q_h^{(r)}(\tau)$ and the maximum number of packets that can be processed per time slot for VNF $V_h^{(r)}$ at the corresponding NFV node.

For VNF $V_h^{(r)}$, the urgent packets with residual lifetime $m = H_r - h + 1$ have the highest priority to be processed, followed by the non-urgent packets whose priority decreases in ascending order of residual lifetime, corresponding to a first-come-first-serve (FCFS) prioritization principle. Let $\hat{S}_{h,m}^{(r)}(\tau)$ be the number of packets with residual lifetime $m \in \mathcal{M}_h^{(r)}$ that are processed at VNF $V_h^{(r)}$ during time slot τ if $z_h^{(r)}(\tau) = 1$, which is calculated as

$$\hat{S}_{h,m}^{(r)}(\tau) = \begin{cases} Q_{h,m}^{(r)}(\tau), & \text{if } m < m_0 \\ \min\left(Q_h^{(r)}(\tau), \left\lfloor \frac{\sum_{n\in\mathcal{N}} x_n^{rh} C_n}{P_h^{(r)}} \right\rfloor\right) - \sum_{m<m_0} Q_{h,m}^{(r)}(\tau), & \text{if } m = m_0 \\ 0, & \text{otherwise} \end{cases} \tag{5.8}$$

where $m_0 \in \mathcal{M}_h^{(r)}$ satisfies

$$\sum_{m\le m_0} Q_{h,m}^{(r)}(\tau) \ge \min\left(Q_h^{(r)}(\tau), \left\lfloor \frac{\sum_{n\in\mathcal{N}} x_n^{rh} C_n}{P_h^{(r)}} \right\rfloor\right) > \sum_{m<m_0} Q_{h,m}^{(r)}(\tau). \tag{5.9}$$

According to the definitions of $\hat{S}_{h,m}^{(r)}(\tau)$ and $S_{h,m}^{(r)}(\tau)$, we have

$$S_{h,m}^{(r)}(\tau) = z_h^{(r)}(\tau) \cdot \hat{S}_{h,m}^{(r)}(\tau). \tag{5.10}$$

5.2 Stochastic VNF Scheduling

5.2.1 Scheduling Problem Formulation

The throughput, $\bar{f}_r$, of deadline-constrained service $r \in \mathcal{R}$ is the difference between the long-term average arrival rate and the long-term average packet dropping rate due to deadline violation, both in packet per time slot, as given by

$$\bar{f}_r = \lim_{\Gamma \to \infty} \frac{1}{\Gamma} \sum_{\tau=0}^{\Gamma-1} \mathbb{E}\left\{ A^{(r)}(\tau) - \sum_{h \in \mathcal{H}_r} D_h^{(r)}(\tau) \right\}, \quad \forall r \in \mathcal{R} \tag{5.11}$$

under the mean rate stable condition for all the physical packet processing queues, represented by

$$\lim_{\Gamma \to \infty} \frac{\mathbb{E}\left\{ \sum_{r \in \mathcal{R}} \sum_{h \in \mathcal{H}_r} q_h^{(r)}(\Gamma) \right\}}{\Gamma} = 0. \tag{5.12}$$

To satisfy the QoS requirement, the throughput of service $r \in \mathcal{R}$ should satisfy

$$\bar{f}_r \geq \bar{a}_r(1 - \varepsilon_r), \quad \forall r \in \mathcal{R}. \tag{5.13}$$

We consider a strictly increasing and concave utility function ϕ of throughput with proportional fairness among services in terms of processing resources, given by

$$\phi(\bar{f}_r) = \log\left(P_r \bar{f}_r\right) \tag{5.14}$$

where $P_r = \sum_{h \in \mathcal{H}_r} P_h^{(r)}$ is the aggregate processing density (in cycle/packet) of service $r \in \mathcal{R}$. In (5.14), the throughput in packet per time slot of service $r \in \mathcal{R}$ (i.e., $\bar{f}_r$) is weighted by P_r, to represent the total throughput in cycle per time slot at all VNFs of the service.

To meet the QoS requirements of all the deadline-constrained services in the presence of unpredictable small-timescale traffic burstiness, we investigate a delay-aware dynamic VNF scheduling problem, which determines which VNF to schedule at each NFV node at each time slot τ, represented by a decision variable set $z(\tau) = \{z_h^{(r)}(\tau), \forall r, \forall h\}$, to maximize a fairness-aware total network utility with QoS guarantee for each service. The VNF scheduling problem is formulated as a stochastic optimization problem, given by

$$\mathbf{P}_\infty : \max_{\{z(\tau),\forall\tau\}} \quad \sum_{r\in\mathcal{R}} \phi(\bar{f}_r)$$
$$\text{s.t.} \quad (5.1), (5.12), (5.13) \tag{5.15}$$

where the objective function and constraints (5.12)–(5.13) involve infinite-horizon expectations, and constraint (5.1) is an instantaneous constraint for each time slot.

5.2.2 Lyapunov Optimization and Problem Transformation

By introducing auxiliary decision variables $\boldsymbol{\chi}(\tau) = \{\chi^{(r)}(\tau) \in [0, A^{(r)}_{max}], \forall r \in \mathcal{R}\}$ for time slot τ, problem $\mathbf{P}_\infty$ is transformed to

$$\mathbf{P}'_\infty : \max_{\{z(\tau),\boldsymbol{\chi}(\tau),\forall\tau\}} \quad \lim_{\Gamma\to\infty} \frac{1}{\Gamma} \sum_{\tau=0}^{\Gamma-1} \mathbb{E}\left\{\sum_{r\in\mathcal{R}} \phi\left(\chi^{(r)}(\tau)\right)\right\} \tag{5.16a}$$
$$\text{s.t.} \quad (5.1), (5.12), (5.13) \tag{5.16b}$$
$$\bar{\chi}_r \le \bar{f}_r, \quad \forall r \in \mathcal{R} \tag{5.16c}$$

where $\bar{\chi}_r = \lim_{\Gamma\to\infty} \frac{1}{\Gamma} \sum_{\tau=0}^{\Gamma-1} \mathbb{E}\left\{\chi^{(r)}(\tau)\right\}$ is the infinite-horizon time average expectations of $\chi^{(r)}(\tau)$ for service $r \in \mathcal{R}$. Problem $\mathbf{P}'_\infty$ is equivalent to problem $\mathbf{P}_\infty$ [2, 3].

To handle the stochastic inequality constraints (5.13) and (5.16c), problem $\mathbf{P}'_\infty$ requires further transformation through introducing equivalent virtual queue stability constraints [3]. Introduce two virtual queues for service $r \in \mathcal{R}$, i.e., $W^{(r)}(\tau)$ and $F^{(r)}(\tau)$, with queue length evolution equations given by

$$W^{(r)}(\tau+1) = \left[W^{(r)}(\tau) - A^{(r)}(\tau) - \sum_{h\in\mathcal{H}_r} S^{(r)}_{h,\mathbb{U}}(\tau)\right]^+ + \bar{a}_r\,(1-\varepsilon_r) + \sum_{h\in\mathcal{H}_r} Q^{(r)}_{h,\mathbb{U}}(\tau) \tag{5.17}$$

$$F^{(r)}(\tau+1) = \left[F^{(r)}(\tau) - A^{(r)}(\tau) - \sum_{h\in\mathcal{H}_r} S^{(r)}_{h,\mathbb{U}}(\tau)\right]^+ + \chi^{(r)}(\tau) + \sum_{h\in\mathcal{H}_r} Q^{(r)}_{h,\mathbb{U}}(\tau) \tag{5.18}$$

where $Q^{(r)}_{h,\mathbb{U}}(\tau)$ and $S^{(r)}_{h,\mathbb{U}}(\tau)$ are the total and processed numbers of urgent packets in the queue of VNF $V^{(r)}_h$ during time slot τ respectively, with $Q^{(r)}_{h,\mathbb{U}}(\tau) = Q^{(r)}_{h,H_r-h+1}(\tau)$ and $S^{(r)}_{h,\mathbb{U}}(\tau) = S^{(r)}_{h,H_r-h+1}(\tau)$. Specifically, from (5.17) we obtain

$\frac{W^{(r)}(\Gamma)-W^{(r)}(0)}{\Gamma} + \frac{1}{\Gamma}\sum_{\tau=0}^{\Gamma-1} A^{(r)}(\tau) \geq \bar{a}_r\,(1-\varepsilon_r) + \frac{1}{\Gamma}\sum_{\tau=0}^{\Gamma-1}\sum_{h\in\mathcal{H}_r} D_h^{(r)}(\tau)$ where $D_h^{(r)}(\tau) = Q_{h,\mathbb{U}}^{(r)}(\tau) - S_{h,\mathbb{U}}^{(r)}(\tau)$. Taking expectations of both sides with $\Gamma \to \infty$ and using $W^{(r)}(0) = 0$, we obtain $\lim_{\Gamma\to\infty} \frac{\mathbb{E}\{W^{(r)}(\Gamma)\}}{\Gamma} + \bar{f}_r \geq \bar{a}_r\,(1-\varepsilon_r)$. Then, the mean rate stability of virtual queue $W^{(r)}(\tau)$, represented by $\lim_{\Gamma\to\infty} \frac{\mathbb{E}\{W^{(r)}(\Gamma)\}}{\Gamma} = 0$, guarantees the stochastic constraint of $\bar{f}_r \geq \bar{a}_r\,(1-\varepsilon_r)$ for service $r \in \mathcal{R}$. Similarly, the mean rate stability of virtual queue $F^{(r)}(\tau)$ enforces the constraint of $\bar{\chi}_r \leq \bar{f}_r$ for service $r \in \mathcal{R}$. Then, problem $\mathbf{P}'_\infty$ is further transformed to

$$\mathbf{P}''_\infty : \max_{\{z(\tau),\chi(\tau),\forall\tau\}} \quad \lim_{\Gamma\to\infty} \frac{1}{\Gamma}\sum_{\tau=0}^{\Gamma-1} \mathbb{E}\left\{\sum_{r\in\mathcal{R}} \phi\left(\chi^{(r)}(\tau)\right)\right\} \tag{5.19a}$$

$$\text{s.t.} \quad (5.1),\ (5.12) \tag{5.19b}$$

$$\lim_{\Gamma\to\infty} \frac{\mathbb{E}\{W^{(r)}(\Gamma)\}}{\Gamma} = \lim_{\Gamma\to\infty} \frac{\mathbb{E}\{F^{(r)}(\Gamma)\}}{\Gamma} = 0, \quad \forall r \in \mathcal{R}. \tag{5.19c}$$

Let $\boldsymbol{q}(\tau) = \{q_h^{(r)}(\tau), \forall r, \forall h\}$, $\boldsymbol{F}(\tau) = \{F^{(r)}(\tau), \forall r\}$ and $\boldsymbol{W}(\tau) = \{W^{(r)}(\tau), \forall r\}$ be the physical and virtual queue vectors at time slot τ. Let $\boldsymbol{\Theta}(\tau) = [\boldsymbol{q}(\tau), \boldsymbol{W}(\tau), \boldsymbol{F}(\tau)]$ be the combined queue vector at time slot τ. Without loss of generality, assume that all queue buffers are infinite and all queues are initially empty at $\tau = 0$. Define Lyapunov function L as a scalar metric of congestion level in the queueing system, given by

$$L\left(\boldsymbol{\Theta}(\tau)\right) = \frac{1}{2}\left\{\sum_{r\in\mathcal{R}}\sum_{h\in\mathcal{H}_r}\left[P_h^{(r)} q_h^{(r)}(\tau)\right]^2 + \sum_{r\in\mathcal{R}}\left[P_r F^{(r)}(\tau)\right]^2 + \sum_{r\in\mathcal{R}}\left[\varphi\frac{W^{(r)}(\tau)}{\bar{a}_r}\right]^2\right\}. \tag{5.20}$$

Note that the delay-aware virtual packet processing queue lengths $\{Q_{h,m}^{(r)}(\tau), \forall r, \forall h, \forall m\}$ are not directly included in the Lyapunov function, since the stability of physical packet processing queues infer the stability of delay-aware virtual packet processing queues. However, the urgent packet queue lengths are implicitly incorporated via virtual queue lengths $W^{(r)}(\tau)$ and $F^{(r)}(\tau)$ according to (5.17) and (5.18). In (5.20), $q_h^{(r)}(\tau)$ is weighted by the processing density $P_h^{(r)}$ of VNF $V_h^{(r)}$, to represent a processing queue backlog in the number of required CPU cycles. The virtual queue length $F^{(r)}(\tau)$ is weighted by the aggregate processing density P_r of service $r \in \mathcal{R}$, since the throughput of each services is weighted by the individual aggregate processing density in the total network utility function according to (5.14). The virtual queue length $W^{(r)}(\tau)$ is normalized by the average number of packet arrivals in each time slot, i.e., $\bar{a}_r$, and then rescaled using a weight $\varphi = [\max_{r\in\mathcal{R}} \bar{a}_r]\left[\max_{r\in\mathcal{R},h\in\mathcal{H}_r} P_h^{(r)}\right]$, to place equal importance on the throughput guarantee for each service regardless of their packet arrival rates and processing densities. A simplified Lyapunov function $L\left(\boldsymbol{\Theta}(\tau)\right) = \frac{1}{2}\left\{\sum_{r\in\mathcal{R}}\sum_{h\in\mathcal{H}_r}\left[P_h^{(r)}q_h^{(r)}(\tau)\right]^2\right\}$ which is unaware of the virtual queue congestion levels corresponds to a classical backpressure algorithm adapted for processing resources. To keep the physical and virtual queues stable by persistently pushing the Lyapunov function $L\left(\boldsymbol{\Theta}(\tau)\right)$ towards a lower congestion level, a one-step conditional Lyapunov drift Δ is introduced, given by

$$\Delta\left(\boldsymbol{\Theta}(\tau)\right) = \mathbb{E}\left\{L\left(\boldsymbol{\Theta}(\tau+1)\right) - L\left(\boldsymbol{\Theta}(\tau)\right)|\boldsymbol{\Theta}(\tau)\right\}. \tag{5.21}$$

Based on the Lyapunov optimization theory, we can decouple the decision variables of problem $\mathbf{P}''_{\infty}$ over time slots and achieve asymptotically optimal total utility with queue stability, by solving an instantaneous problem for each time slot [3–6]. At time slot τ, the upper bound of a conditional Lyapunov drift-plus-penalty (or drift-minus-utility) term defined as $\Delta\left(\boldsymbol{\Theta}(\tau)\right) - \vartheta\mathbb{E}\left\{\sum_{r\in\mathcal{R}}\phi(\chi^{(r)}(\tau))|\boldsymbol{\Theta}(\tau)\right\}$ is minimized, subject to the instantaneous constraint (5.1). Here, ϑ (> 0) is a utility importance parameter that balances the importance between utility maximization and queue backlog reduction. The upper bound of the conditional Lyapunov drift-plus-penalty is given in Lemma 5.1.

Lemma 5.1 *Regardless of the randomness in packet arrivals and VNF scheduling decisions, the conditional Lyapunov drift-plus-penalty at time slot τ, has an upper bound, given by*

$$\begin{aligned}
&\Delta\left(\boldsymbol{\Theta}(\tau)\right) - \vartheta\mathbb{E}\left\{\sum_{r\in\mathcal{R}}\phi\left(\chi^{(r)}(\tau)\right)|\boldsymbol{\Theta}(\tau)\right\}\\
&\le \sum_{r\in\mathcal{R}}\mathsf{B}_r + \sum_{r\in\mathcal{R}}\mathbb{E}\left\{\left[\left(P_1^{(r)}\right)^2 q_1^{(r)}(\tau) - \left(\frac{\varphi}{\bar{a}_r}\right)^2 W^{(r)}(\tau)\right.\right.\\
&\quad\left.\left. - \left(P_r\right)^2 F^{(r)}(\tau)\right] A^{(r)}(\tau)|\boldsymbol{\Theta}(\tau)\right\}\\
&\quad + \sum_{r\in\mathcal{R}}\mathbb{E}\left\{\frac{\varphi^2\left(1-\varepsilon_r\right)}{\bar{a}_r}W^{(r)}(\tau) + \left[\left(\frac{\varphi}{\bar{a}_r}\right)^2 W^{(r)}(\tau)\right.\right.\\
&\quad\left.\left. + \left(P_r\right)^2 F^{(r)}(\tau)\right]\sum_{h\in\mathcal{H}_r} Q_{h,\mathbb{U}}^{(r)}(\tau)|\boldsymbol{\Theta}(\tau)\right\}\\
&\quad - \sum_{r\in\mathcal{R}}\mathbb{E}\left\{\Phi_1^{(r)}(\tau)|\boldsymbol{\Theta}(\tau)\right\} - \sum_{r\in\mathcal{R}}\mathbb{E}\left\{\Phi_2^{(r)}(\tau)|\boldsymbol{\Theta}(\tau)\right\}
\end{aligned} \tag{5.22}$$

where B_r is a constant given in Appendix B, and $\Phi_1^{(r)}(\tau)$, $\Phi_2^{(r)}(\tau)$ are given by

$$\Phi_1^{(r)}(\tau) = \vartheta \cdot \phi\left(\chi^{(r)}(\tau)\right) - (P_r)^2 \left[F^{(r)}(\tau) + \sum_{h \in \mathcal{H}_r} Q_{h,\mathbb{U}}^{(r)}(\tau) \right] \chi^{(r)}(\tau), \quad \forall r \in \mathcal{R} \tag{5.23}$$

$$\begin{aligned} \Phi_2^{(r)}(\tau) = & \sum_{h \in \mathcal{H}_r} \left(P_h^{(r)}\right)^2 q_h^{(r)}(\tau) \left[S_h^{(r)}(\tau) - S_{h-1}^{(r)}(\tau) \mathbb{1}_{\{h>1\}} \right] \\ & + \sum_{h \in \mathcal{H}_r} \left[\left(\frac{\varphi}{\bar{a}_r}\right)^2 W^{(r)}(\tau) + (P_r)^2 F^{(r)}(\tau) \right. \\ & \left. - \left(P_h^{(r)}\right)^2 q_h^{(r)}(\tau) \right] S_{h,\mathbb{U}}^{(r)}(\tau), \ \forall r \in \mathcal{R}. \end{aligned} \tag{5.24}$$

Proof See Appendix B.

Since only the last two terms of the upper bound in (5.22) are related to the decision variables, we can formulate an instantaneous problem at time slot τ as

$$\begin{aligned} \mathbf{P}_\tau : \ & \max_{z(\tau), \boldsymbol{\chi}(\tau)} \quad \sum_{r \in \mathcal{R}} \mathbb{E}\left\{\Phi_1^{(r)}(\tau) | \boldsymbol{\Theta}(\tau)\right\} + \sum_{r \in \mathcal{R}} \mathbb{E}\left\{\Phi_2^{(r)}(\tau) | \boldsymbol{\Theta}(\tau)\right\} \\ & \text{s.t.} \quad \sum_{(r,h) \in \mathcal{V}_n} z_h^{(r)}(\tau) \leq 1, \quad \forall n \in \mathcal{N}. \end{aligned} \tag{5.25}$$

In problem $\mathbf{P}_\tau$, we have two groups of decision variables, $\boldsymbol{\chi}(\tau)$ and $z(\tau)$, which are separable in both the objective function and constraints. Thus, problem $\mathbf{P}_\tau$ is equivalent to two sub-problems, $\mathbf{P}_{\tau,1}$ and $\mathbf{P}_{\tau,2}$, given by

$$\mathbf{P}_{\tau,1} : \max_{\boldsymbol{\chi}(\tau)} \ \sum_{r \in \mathcal{R}} \mathbb{E}\left\{\Phi_1^{(r)}(\tau) | \boldsymbol{\Theta}(\tau)\right\} \tag{5.26}$$

$$\begin{aligned} \mathbf{P}_{\tau,2} : \ & \max_{z(\tau)} \ \sum_{r \in \mathcal{R}} \mathbb{E}\left\{\Phi_2^{(r)}(\tau) | \boldsymbol{\Theta}(\tau)\right\} \\ & \text{s.t.} \quad \sum_{(r,h) \in \mathcal{V}_n} z_h^{(r)}(\tau) \leq 1, \quad \forall n \in \mathcal{N}. \end{aligned} \tag{5.27}$$

Using the concept of opportunistically maximizing an expectation, the objective function of sub-problems $\mathbf{P}_{\tau,1}$ and $\mathbf{P}_{\tau,2}$ can be maximized by maximizing $\sum_{r \in \mathcal{R}} \Phi_1^{(r)}(\tau)$ or $\sum_{r \in \mathcal{R}} \Phi_2^{(r)}(\tau)$ given the observed values of $\boldsymbol{\Theta}(\tau)$ under the corresponding constraints [3].

5.3 Online Distributed Algorithm

We design an online distributed VNF scheduling algorithm to solve sub-problems $\mathbf{P}_{\tau,1}$ and $\mathbf{P}_{\tau,2}$ for each time slot, with a flowchart given in Fig. 5.5.

5.3.1 *Auxiliary Variable Decision*

Sub-problem $\mathbf{P}_{\tau,1}$ is separable among services. For service $r \in \mathcal{R}$, since only virtual queue length $F^{(r)}(\tau)$ and urgent packet queue lengths $\{Q_{h,\mathbb{U}}^{(r)}(\tau), \forall h \in \mathcal{H}_r\}$ among $\mathbf{\Theta}(\tau)$ appear in $\Phi_1^{(r)}(\tau)$, the optimal value of $\chi^{(r)}(\tau)$ can be derived by observing $F^{(r)}(\tau)$ and $\{Q_{h,\mathbb{U}}^{(r)}(\tau), \forall h \in \mathcal{H}_r\}$, and solving an optimization problem given by

$$\mathbf{P}_{\tau,1}^{(r)}: \max_{\chi^{(r)}(\tau)\in\left[0, A_{max}^{(r)}\right]} \vartheta \cdot \phi(\chi^{(r)}(\tau)) - (P_r)^2 \left[F^{(r)}(\tau) + \sum_{h\in\mathcal{H}_r} Q_{h,\mathbb{U}}^{(r)}(\tau) \right] \chi^{(r)}(\tau). \tag{5.28}$$

The optimal solution of problem $\mathbf{P}_{\tau,1}^{(r)}$, denoted by $\chi^{(r)*}(\tau)$, is derived by differentiating the objective function with respect to $\chi^{(r)}(\tau)$, given by

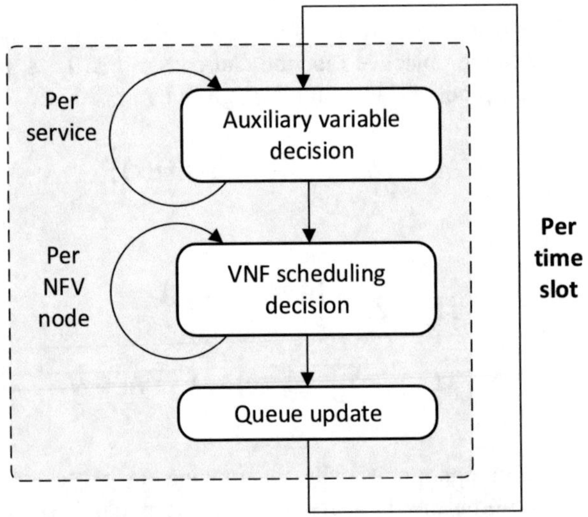

Fig. 5.5 Flowchart of the online distributed VNF scheduling algorithm

$$\chi^{(r)*}(\tau) = \begin{cases} A_{max}^{(r)}, & \text{if } (P_r)^2\left[F^{(r)}(\tau) + \sum_{h\in\mathcal{H}_r} Q_{h,\mathbb{U}}^{(r)}(\tau)\right] \le \frac{\vartheta}{A_{max}^{(r)}} \\ \frac{\vartheta}{(P_r)^2\left[F^{(r)}(\tau)+\sum_{h\in\mathcal{H}_r} Q_{h,\mathbb{U}}^{(r)}(\tau)\right]}, & \text{otherwise.} \end{cases} \tag{5.29}$$

5.3.2 VNF Scheduling Decision

Let $S_{h,\mathbb{N}}^{(r)}(\tau)$ denote the number of non-urgent packets processed at VNF $V_h^{(r)}$ of service $r \in \mathcal{R}$ during time slot τ, with $S_{h,\mathbb{N}}^{(r)}(\tau) = S_h^{(r)}(\tau) - S_{h,\mathbb{U}}^{(r)}(\tau)$. Then, by rewriting $\sum_{r\in\mathcal{R}} \Phi_2^{(r)}(\tau)$ as a summation over $n \in \mathcal{N}$, sub-problem $\mathbf{P}_{\tau,2}$ can be further decomposed into NFV node level sub-problems. For NFV node $n \in \mathcal{N}$, the optimal VNF scheduling decisions $\{z_h^{(r)}(\tau), \forall(r,h) \in \mathcal{V}_n\}$ can be obtained by maximizing a linearly weighted summation of $S_{h,\mathbb{N}}^{(r)}(\tau)$ and $S_{h,\mathbb{U}}^{(r)}(\tau)$, under the single VNF scheduling constraint, given by

$$\mathbf{P}_{\tau,2}^{(n)}: \max_{\left\{z_h^{(r)}(\tau)\middle|(r,h)\in\mathcal{V}_n,\sum_{n\in\mathcal{V}_n} z_h^{(r)}(\tau)\le 1\right\}} \sum_{(r,h)\in\mathcal{V}_n} \left[\omega_{h,\mathbb{U}}^{(r)}(\tau)S_{h,\mathbb{U}}^{(r)}(\tau) + \omega_{h,\mathbb{N}}^{(r)}(\tau)S_{h,\mathbb{N}}^{(r)}(\tau)\right] \tag{5.30}$$

where $\omega_{h,\mathbb{U}}^{(r)}(\tau)$ and $\omega_{h,\mathbb{N}}^{(r)}(\tau)$ are the adaptive scheduling weights for one urgent packet and for one non-urgent packet at VNF $V_h^{(r)}$ during time slot τ respectively, given by

$$\omega_{h,\mathbb{U}}^{(r)}(\tau) = \left(\frac{\varphi}{\bar{a}_r}\right)^2 W^{(r)}(\tau) + (P_r)^2 F^{(r)}(\tau) - \left(P_{h+1}^{(r)}\right)^2 q_{h+1}^{(r)}(\tau) \tag{5.31}$$

$$\omega_{h,\mathbb{N}}^{(r)}(\tau) = \left(P_h^{(r)}\right)^2 q_h^{(r)}(\tau) - \left(P_{h+1}^{(r)}\right)^2 q_{h+1}^{(r)}(\tau). \tag{5.32}$$

Here, we have $P_{H_r+1}^{(r)} \equiv q_{H_r+1}^{(r)} \equiv 0$ for $h = H_r$. The scheduling weight for one urgent packet at VNF $V_h^{(r)}$, i.e., $\omega_{h,\mathbb{U}}^{(r)}(\tau)$, corresponds to the difference between the weighted virtual queue lengths of service $r \in \mathcal{R}$ and the weighted physical packet processing queue length at downstream VNF $V_{h+1}^{(r)}$, while the scheduling weight for one non-urgent packet at VNF $V_h^{(r)}$, i.e., $\omega_{h,\mathbb{N}}^{(r)}(\tau)$, corresponds to the weighted differential backlogs between VNF $V_h^{(r)}$ and downstream VNF $V_{h+1}^{(r)}$. Through such a differentiation between urgent and non-urgent packets, packet urgency and throughput performance are incorporated in VNF scheduling beyond the classical backpressure scheduling policy. A temporary greater congestion level in the virtual queues indicates less satisfaction or even violation of the service throughput requirement, resulting in a larger scheduling weight for each urgent packet in (5.31). A greater congestion level at the downstream VNF discourages packet processing at the upstream VNF to avoid further worsening the congestion situation, through reducing the packet scheduling weights for both urgent and non-urgent packets at the upstream VNF. For the classical backpressure algorithm adapted for processing resources, no differentiation is considered between urgent and non-urgent packets, and all packets are treated as non-urgent packets with the same scheduling weight in (5.32).

If VNF $V_h^{(r)}$ is scheduled during time slot τ, we have $S_{h,\mathbb{U}}^{(r)}(\tau) = \hat{S}_{h,H_r-h+1}^{(r)}(\tau)$ and $S_{h,\mathbb{N}}^{(r)}(\tau) = \sum_{m>H_r-h+1} \hat{S}_{h,m}^{(r)}(\tau)$ according to (5.8). Define a VNF scheduling weight $\omega_h^{(r)}(\tau)$ as a summation of the total packet scheduling weights for all the urgent and non-urgent packets that are processed at VNF $V_h^{(r)}$ if VNF $V_h^{(r)}$ is scheduled during time slot τ, given by

$$\omega_h^{(r)}(\tau) = \omega_{h,\mathbb{U}}^{(r)}(\tau)\hat{S}_{h,H_r-h+1}^{(r)}(\tau) + \omega_{h,\mathbb{N}}^{(r)}(\tau) \sum_{m>H_r-h+1} \hat{S}_{h,m}^{(r)}(\tau),$$
$$\forall r \in \mathcal{R},\ h \in \mathcal{H}_r. \tag{5.33}$$

Then, the VNF with the largest VNF scheduling weight is greedily scheduled at each NFV node, and the optimal solutions for problem $\mathbf{P}_{\tau,2}^{(n)}$ associated with NFV node $n \in \mathcal{N}$ are given by

$$z_h^{(r)*}(\tau) = \mathbb{1}_{\left\{(r,h)=\arg\max_{(r,h)\in\mathcal{V}_n} \omega_h^{(r)}(\tau)\right\}}, \quad \forall (r,h) \in \mathcal{V}_n. \tag{5.34}$$

For a service, a temporal throughput degradation below the minimum requirement results in a higher congestion level in the virtual queues and more urgent packets in the packet processing queues, which in turn increases the VNF scheduling weights for all VNFs in the service. In this way, the VNFs have more chances to be scheduled, leading to an improvement in the throughput.

For the classical backpressure algorithm adapted for processing resources without packet urgency awareness, the VNF scheduling weight for VNF $V_h^{(r)}$ at time slot τ is simplified as $\omega_h^{(r)}(\tau) = \omega_{h,\mathbb{N}}^{(r)}(\tau) \min\left(Q_h^{(r)}(\tau), \left\lfloor \frac{\sum_{n\in\mathcal{N}} x_n^{rh} C_n}{P_h^{(r)}} \right\rfloor\right)$.

5.3.3 Queue Updates

Combining all the decisions for time slot τ, the queue backlogs for time slot $\tau + 1$ including the physical packet processing queue lengths, $\{q_h^{(r)}(\tau + 1), \forall r, h\}$, the virtual packet processing queue lengths, $\{Q_{h,m}^{(r)}(\tau + 1), \forall r, h, m\}$, and the service-level virtual queue lengths, $\{W^{(r)}(\tau+1), F^{(r)}(\tau+1), \forall r\}$, are updated according to (5.2), (5.4), (5.17) and (5.18).

5.3.4 *Performance Optimality*

The online VNF scheduling algorithm achieves $O(\frac{1}{\vartheta})$ near-optimal total utility, with the optimality gap decreasing with ϑ, and results in linearly increasing total queue backlogs with the increase of ϑ, demonstrating an $\left[O(\frac{1}{\vartheta}), O(\vartheta)\right]$ utility-backlog trade-off [3].

5.4 Performance Evaluation

We consider two virtual network topologies, both with 6 services of given VNF placement at NFV nodes, as shown in Table 5.1, where n_i denotes the i-th NFV node. The services in topology 1 traverse through different virtual paths in a network of 9 NFV nodes, while all the services in topology 2 share a common virtual path through 4 NFV nodes. The packet E2E deadline of each service is set as 10 ms. We assume that the maximum packet dropping ratios for different services are the same, denoted by ε. By default, ε is set as 10^{-3}. We use 6 real-world stationary traffic traces with packet timestamp information [7, 8]. The average packet arrival rates of the 6 traffic traces are 17915, 25627, 33038, 51182, 47810, 67912 in packet/s respectively. We consider two traffic sets for the services, as given in Table 5.2, where the number inside the bracket indicates the processing density in kilo-cycle per packet (i.e., $Kcpp$). For example, in traffic set 1, we use traffic trace 1 for service 1, with a processing density of $1Kcpp$ for each VNF in the service [9]. In traffic set 1, the services have different traffic traces and the same processing density. In traffic set 2, each traffic trace is used for two services with different processing densities. We use topology 1 and traffic set 1 as the default simulation setting. The time slot length T is set as 1 ms by default. The total processing resource budget (in cycle per time slot) at NFV node $n \in \mathcal{N}$ is proportional to the average processing resource demand of all the VNFs placed at the NFV node, given by

Table 5.1 Simulation settings for virtual network topology

Topology	Services
1	Service 1: $n_1 \to n_2 \to n_3 \to n_7$; Service 2: $n_4 \to n_1 \to n_7 \to n_2$
	Service 3: $n_8 \to n_5 \to n_2 \to n_6$; Service 4: $n_3 \to n_9 \to n_6 \to n_1$
	Service 5: $n_5 \to n_6 \to n_4 \to n_3$; Service 6: $n_9 \to n_7 \to n_8 \to n_2$
2	All services: $n_1 \to n_2 \to n_3 \to n_4$

Table 5.2 Traffic sets for VNF scheduling simulation

Set	Service 1	Service 2	Service 3	Service 4	Service 5	Service 6
1	Trace 1 (1)	Trace 2 (1)	Trace 3 (1)	Trace 4 (1)	Trace 5 (1)	Trace 6 (1)
2	Trace 1 (1)	Trace 1 (4)	Trace 2 (1)	Trace 2 (4)	Trace 3 (1)	Trace 3 (4)

Table 5.3 Default parameters in VNF scheduling

Parameter	Definition	Value
M_r	Packet E2E deadline	10 ms
T	Time slot length	1 ms
ε	Maximum packet dropping ratio	10^{-3}
$P_h^{(r)}$	Processing density	1 Kcpp
ϑ	Utility importance parameter	10^5
Γ	Total number of time slots	10^5

$$C_n = \rho \sum_{(r,h)\in\mathcal{V}_n} \bar{a}_r P_h^{(r)}, \quad \forall n \in \mathcal{N} \tag{5.35}$$

where ρ is referred to as the resource overprovisioning ratio. The utility importance parameter, ϑ, is set as 10^5 by default. With a certain simulation setting, let the VNF scheduling algorithm run for $\Gamma = 10^5$ time slots. The performance metrics such as the throughput $\bar{f}_r$ in (5.11) and the average E2E delay $\bar{d}_r$ in (5.6) are calculated based on the average over the Γ time slots. All the default parameters for performance evaluation are summarized in Table 5.3.

With the basic VNF scheduling algorithm, the utility-backlog trade-off is investigated by increasing the utility importance parameter ϑ from 1 to 200000. The resource overprovisioning ratio is set as $\rho = 3$. We also examine the impact of QoS constraints on the utility-backlog trade-off, by setting $\varepsilon = 10^{-3}, 0.5, 0.8, 1$ to represent different levels of relaxation on the QoS constraints. For $\varepsilon = 1$, it corresponds to a utility maximization problem without explicit QoS constraints for each service. Figure 5.6a shows the total utility with the increase of ϑ at different values of ε. In the figure, the "$x\%$ utility" represents the total utility when the timely delivery ratio of each service is $x\%$. With the increase of ϑ, the total utility with $\varepsilon = 10^{-3}$ is stable, which is slightly beyond the 99% utility but does not reach the 99.9% utility, inferring that a resource overprovisioning ratio of $\rho = 3$ is not sufficient for such a strict QoS requirement under the given simulation settings. With a relaxed QoS constraint, i.e., $\varepsilon \in \{0.5, 0.8, 1\}$, the total utility gradually increases and gets closer to the 99% utility with the increase of ϑ. We notice that the achieved total utility without explicit QoS constraints (i.e., $\varepsilon = 1$) at $\vartheta = 1$ is close to the 95% utility. However, for each value of ε, we see an linear increase in the average total backlogs (i.e., average of the total actual and virtual queue lengths over time slots) with the increase of ϑ in Fig. 5.6b, demonstrating an $\left[O(\frac{1}{\vartheta}), O(\vartheta)\right]$ utility-backlog trade-off with the increase of ϑ. We also observe that more utility is achieved with a more strict QoS constraint for a certain value of ϑ, at a cost of a greater congestion level. The average total backlogs with $\varepsilon = 10^{-3}$ is even two orders of magnitude

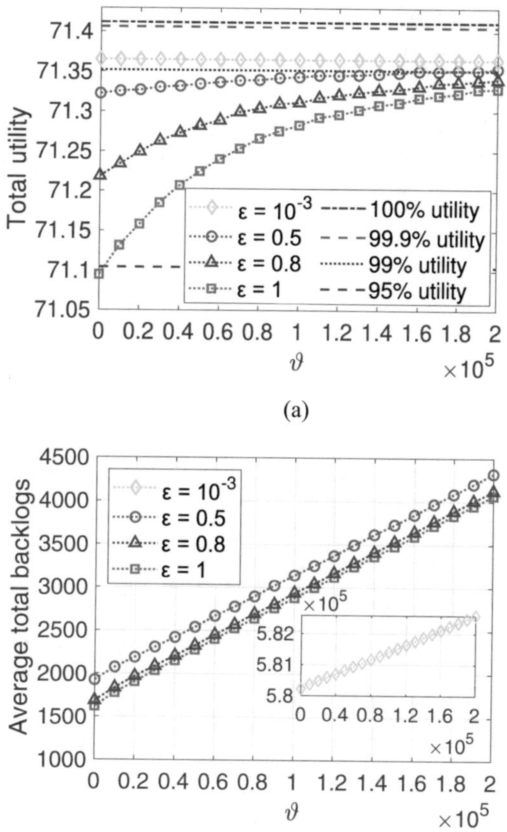

Fig. 5.6 Trade-off between total utility and average total backlogs with respect to ϑ ($\rho = 3$). (**a**) Total utility. (**b**) Average total backlogs

higher than that with relaxed QoS constraints, since the QoS constraint violation with $\varepsilon = 10^{-3}$ results in unstable virtual queue $W^{(r)}(\tau)$ with consistently increasing virtual queue length over time. With a smaller value of ε, the virtual queue length $W^{(r)}(\tau)$ grows more aggressively with the same number of packet dropping due to the term $\bar{a}_r\,(1-\varepsilon_r)$ in (5.17), thus imposing more scheduling weight on the urgent packets according to (5.31). In this way, packet urgency plays a more important role in VNF scheduling, resulting in less packet droppings due to expiry.

Figure 5.7 illustrates the performance comparison between the VNF scheduling algorithm and two benchmark algorithms in terms of the individual timely delivery ratios of different services. We use a simulation setting with topology 2 and traffic set 2, to evaluate the impact of packet arrival rate and processing density on the individual performance of each service with the increase of resource availability. The "GPS-Average" benchmark algorithm corresponds to a GPS resource allocation scheme under the unrealistic assumption of infinitely divisible resources, where VNF $V_h^{(r)}$ enjoys as if a dedicated virtual CPU with a minimum processing rate of

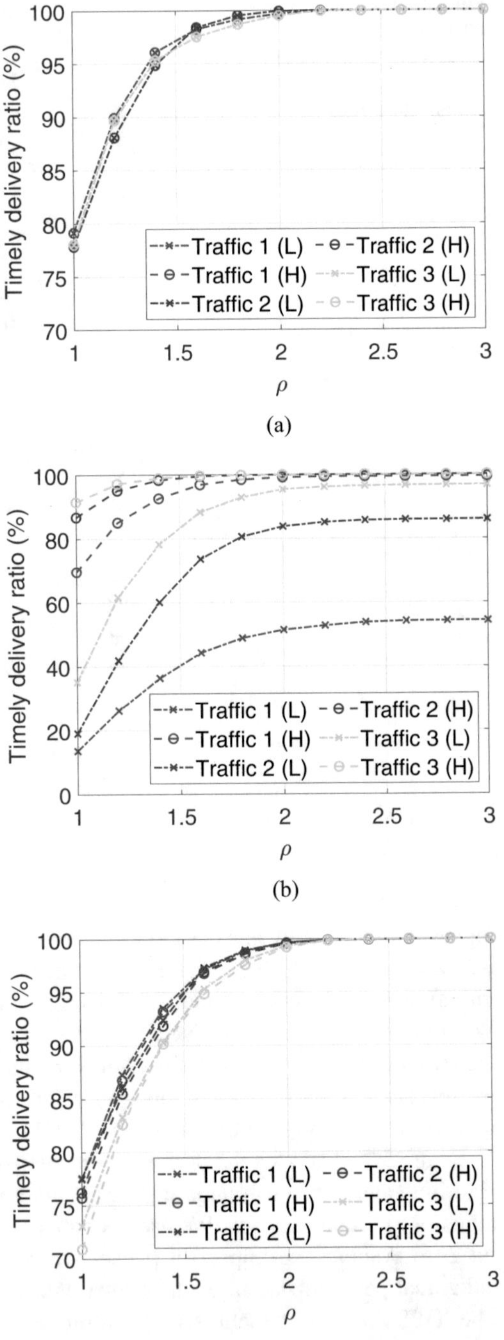

Fig. 5.7 Performance comparison between the VNF scheduling algorithm and benchmarks. (**a**) GPS-average. (**b**) Backpressure-P. (**c**) Proposed

$\rho \bar{a}_r P_h^{(r)}$ in cycle per time slot. The virtual CPUs can be scheduled simultaneously at each NFV node, with multiplexing among each other. We see that the timely delivery ratios of services with the same traffic trace and different processing densities overlap with each other, and the timely delivery ratios of services with different traffic traces are close to each other. The "Backpressure-P" benchmark algorithm corresponds to the classical backpressure algorithm adapted for processing resources, in which the differential backlogs in number of required CPU cycles is used as the VNF scheduling weight, and no virtual queues are introduced for individual QoS guarantee. We see a significant performance degradation for low-density and low-rate services. By contrast, the delay-aware VNF scheduling algorithm takes equal importance of the QoS requirement of each service, and achieves similar timely delivery ratios for each service, regardless of the difference in the packet arrival rate and processing density. Moreover, the performance of the delay-aware VNF scheduling algorithm is comparable to that of the "GPS-Average" algorithm, although the delay-aware VNF scheduling algorithm operates in a time-slotted manner with $T = 1$ ms under the constraint that at most one VNF can be scheduled at each NFV node during each time slot.

In the VNF scheduling algorithm, we have assumed that a packet can be processed by at most one VNF in a chain during one time slot. We refer to this assumption as no packet rushing assumption, as a packet is allowed to "rush" to downstream VNFs for further processing if such an assumption is removed. Consider an extreme case under the assumption of no packet rushing. For a service with H_r VNFs in the chain, a packet experiences at least an E2E delay of $H_r T$ even if there is no packet queueing at all the VNFs, where T is the time slot length. Such a delay overhead is referred to as the worst-case E2E delay overhead, which is non-negligible for a realistic time slot length such as 1 ms and a strict E2E delay requirement such as 10 ms. However, if packet rushing is allowed, when a scheduled VNF $V_h^{(r)}$ $(h > 1)$ is unsaturated for time slot τ, i.e., there are residual CPU cycles before the end of the time slot after all the packets in its queue are processed, corresponding to the condition of $S_h^{(r)}(\tau) = Q_h^{(r)}(\tau) < \left\lfloor \frac{\sum_{n \in N} x_n^{rh} C_n}{P_h^{(r)}} \right\rfloor$, some packets processed by upstream VNF $V_{h-1}^{(r)}$ during time slot τ can be further processed by VNF $V_h^{(r)}$ using the residual CPU cycles during the same time slot, hence enhancing resource utilization and reducing packet E2E delay. A modified VNF scheduling algorithm corrected by the packet rushing effect is given in Appendix C. Next, we evaluate how packet rushing can affect the performance of VNF scheduling.

Figure 5.8 shows comparison of three performance metrics between the basic VNF scheduling algorithm without packet rushing and the modified VNF scheduling algorithm with packet rushing (denoted by "Basic" and "Rush" respectively), including the total throughput (in packet/s) of all services, the average E2E delay of different services, and the average total backlogs. We see an improvement in all the three performance metrics with the increase of resource availability (indicated by ρ) in both cases without and with packet rushing. As illustrated in Fig. 5.8a, more

Fig. 5.8 Performance comparison without and with packet rushing with the increase of resource availability. (**a**) Total throughput (packet/s). (**b**) Average E2E delay (ms). (**c**) Average total backlogs

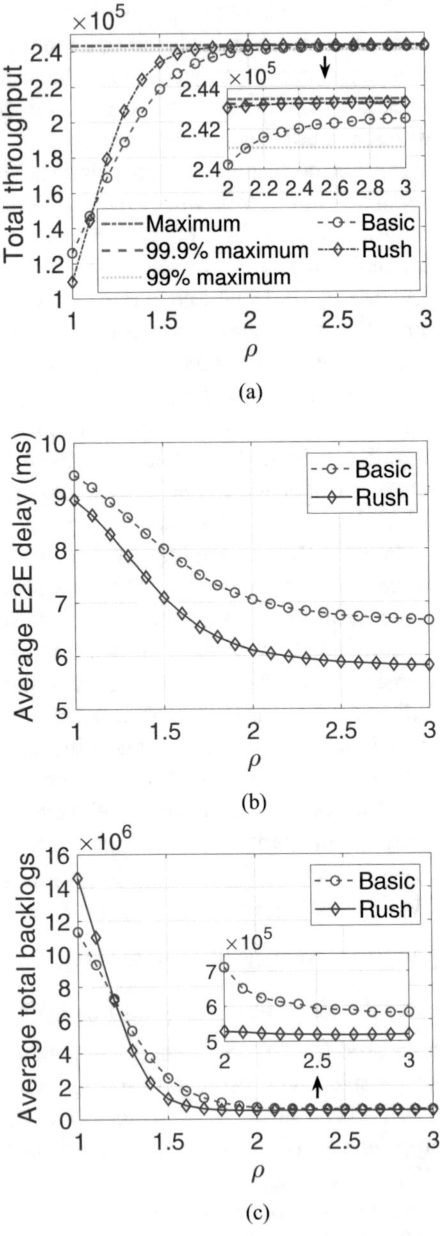

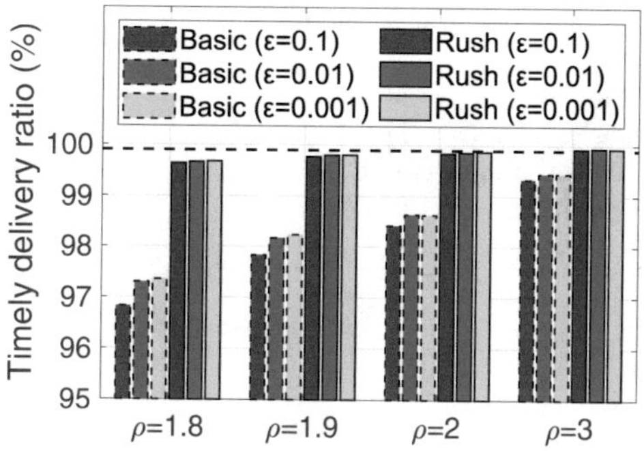

Fig. 5.9 Average timely delivery ratio with different QoS constraints

packets are timely delivered to the egress edge switch within E2E deadline by taking advantage of packet rushing when there are sufficient resources in the network, i.e., when ρ is greater than a certain value around 1.2. The total throughput achieved by the basic and modified algorithms approach 99% and 99.9% of the maximum value at a resource overprovisioning ratio around 2.1, respectively. However, the strict QoS requirement with $\varepsilon = 10^{-3}$ is difficult to be satisfied without packet rushing, even with further increase of ρ beyond 2.1. We also observe that the modified algorithm with packet rushing cannot outperform the basic algorithm in terms of the total throughput when the resources are limited, e.g., $\rho = 1$. The reason is that the resources allocated to the packets with residual lifetime $1 \leq m < H_r - h + 1$ at VNF $V_h^{(r)}$ have high chances to be eventually wasted due to limited packet rushing opportunity at low resource availability, since such packets can be successfully delivered only by taking advantage of packet rushing. Figure 5.8b shows that the average E2E delay of the services is reduced with packet rushing. With the increase of ρ, the gap between the average E2E delay achieved by the basic and modified algorithms increases to around 1 ms. Figure 5.8c shows that the average total backlogs are reduced with more resources, and the reduction is more significant with packet rushing. With the increase of ρ, the QoS performance gradually approaches the QoS requirement as illustrated in Fig. 5.8a, resulting in a reduced congestion level in the virtual queues. Since packet rushing enhances the QoS performance if resources are sufficient, the virtual queues become even less congested with packet rushing. The physical packet processing queues also become less congested with packet rushing, since the packets can reach the egress edge switch faster on average.

Figure 5.9 shows the average timely delivery ratio of different services with different QoS constraints ($\varepsilon = 10^{-1}, 10^{-2}, 10^{-3}$) at given values of ρ for both cases without and with packet rushing. With the same QoS constraint, we observe an improvement in the average timely delivery ratio with the increase of ρ and with packet rushing, which is consistent with the results shown in Fig. 5.8. With packet rushing, the difference between the achieved average timely delivery ratios with different QoS constraints at a certain value of ρ is less significant, since a large portion of urgent packets that should have been dropped under the worst-case assumption of no packet rushing can rush to the egress edge switch before expiry,

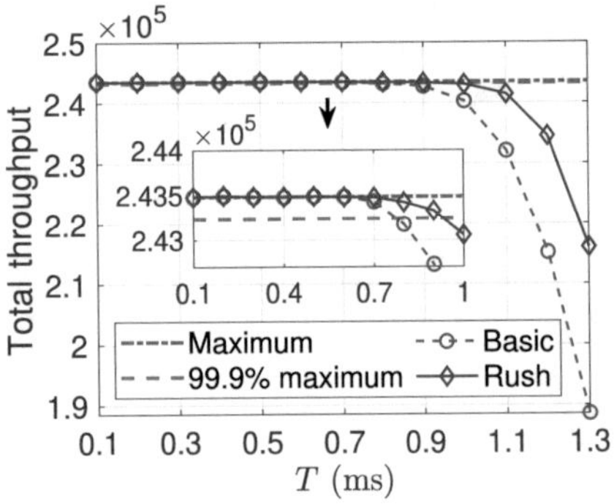

(a)

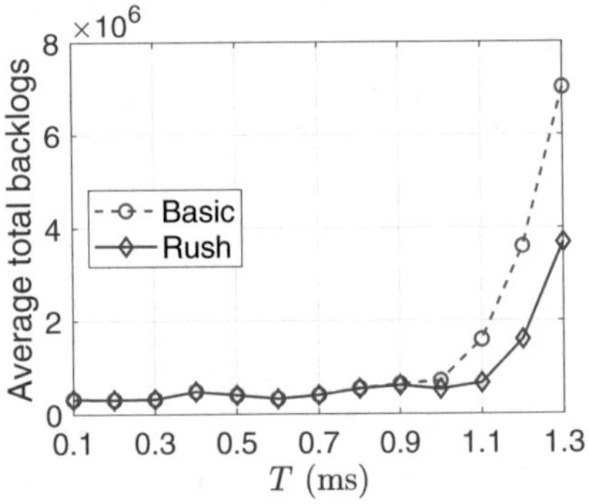

(b)

Fig. 5.10 Performance of the delay-aware VNF scheduling algorithm at different time slot length T ($\rho = 2$). (**a**) Total throughput (packet/s). (**b**) Average total backlogs

and the achieved average timely delivery ratio approaches 1 at the given values of ρ regardless of the QoS constraint.

To see the impact of time slot length (T) on VNF scheduling performance, we evaluate four performance metrics (including the total throughput, the average total backlogs, the average E2E delay, and the number of context switches per second) at $\rho = 2$, by increasing the time slot length T from 0.1 ms to 1.3 ms. Due to space limit, the results are shown in two figures including Figures 5.10 and 5.11. As illustrated in Fig. 5.10a, when the algorithm operates with an extremely small time slot length, e.g., 0.1 ms, almost no packets are dropped due to E2E delay violation, resulting in a total throughput approaching 100% of the maximum value. The total throughput remains high until T increases to around 0.7 ms, and then degrades significantly with further increase of T. Even if packet rushing is allowed, the QoS violation is significant if T is too large. Correspondingly, the average total backlogs first increase very slowly and then increase sharply due to significant QoS degradation with the increase of T, as illustrated in Fig. 5.10b. With the increase of T, the worst-case E2E delay overhead, i.e., $H_r T$ for service $r \in \mathcal{R}$, becomes more significant, which cannot be fully compensated by the delay reduction benefit from packet rushing, resulting in an almost linear increasing trend in the average E2E delay for both cases without and with packet rushing, as illustrated in Fig. 5.11a. The

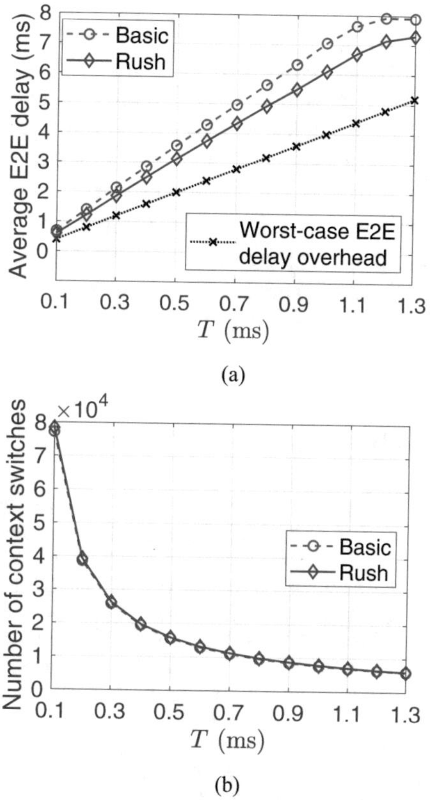

Fig. 5.11 Performance of the delay-aware VNF scheduling algorithm at different time slot length T ($\rho = 2$). (**a**) Average E2E delay (ms). (**b**) No. of context switches per second

former three performance metrics are all improved with finer granularity of time slot length, at a cost of more switching overhead per second which is nearly inversely proportional to T, as illustrated in Fig. 5.11b. Hence, the time slot length should not be too small to avoid significant switching overhead. Moreover, if T is too small, the assumption of negligible transmission and propagation delay over the virtual links between consecutive VNFs is non-realistic, and the complexity of the VNF scheduling algorithm grows due to increased size of the packet residual lifetime set at VNF $V_h^{(r)}$, i.e., $|\mathcal{M}_h^{(r)}|$ or $|\tilde{\mathcal{M}}_h^{(r)}|$.

5.5 Summary

In this chapter, we study a delay-aware VNF scheduling problem for deadline-constrained services in a network slicing scenario, to achieve utility maximization in the presence of small-timescale traffic dynamics, while satisfying the QoS requirement in terms of delay violation ratio. An online distributed VNF scheduling algorithm based on a delay-aware virtual queueing model is presented. The differential backlogs, throughput performance and packet urgency are taken into consideration for VNF scheduling. Simulation results demonstrate an $\left[O(\frac{1}{\vartheta}), O(\vartheta)\right]$ utility-backlog trade-off with utility importance parameter ϑ. The effectiveness of virtual queues is verified through a significant QoS performance gap between the delay-aware VNF scheduling algorithm and the "Backpressure-P" benchmark. Even though the delay-aware VNF scheduling algorithm operates in a time granularity of $100\mu s$ to ms, the performance gap with a GPS scheme under the assumption of infinitely divisible resources is not significant. A performance improvement is observed with packet rushing at a sufficiently high resource availability especially in terms of average E2E delay, and the impact of time slot length on VNF scheduling performance is also evaluated.

References

1. Qu, K., Zhuang, W., Ye, Q., Shen, X., Li, X., Rao, J.: Dynamic flow migration for embedded services in SDN/NFV-enabled 5G core networks. IEEE Trans. Commun. **68**(4), 2394–2408 (2020)
2. Gu, L., Zeng, D., Tao, S., Guo, S., Jin, H., Zomaya, A.Y., Zhuang, W.: Fairness-aware dynamic rate control and flow scheduling for network utility maximization in network service chain. IEEE J. Sel. Areas Commun. **37**(5), 1059–1071 (2019)
3. Neely, M.: Stochastic Network Optimization with Application to Communication and Queueing Systems. Morgan & Claypool Publishers (2010)
4. Lu, P., Sun, Q., Wu, K., Zhu, Z.: Distributed online hybrid cloud management for profit-driven multimedia cloud computing. IEEE Trans. Multimedia **17**(8), 1297–1308 (2015)
5. Zhou, Z., Liu, F., Zou, R., Liu, J., Xu, H., Jin, H.: Carbon-aware online control of geo-distributed cloud services. IEEE Trans. Parallel Distrib. Syst. **27**(9), 2506–2519 (2016)
6. Li, S., Zhou, Y., Jiao, L., Yan, X., Wang, X., et al.: Towards operational cost minimization in hybrid clouds for dynamic resource provisioning with delay-aware optimization. IEEE Trans. Serv. Comput. **8**(3), 398–409 (2015)
7. MAWI Working Group Traffic Archive (2021). http://mawi.wide.ad.jp/mawi/. Accessed 14 July 2020
8. Qu, K., Zhuang, W., Ye, Q., Shen, X., Li, X., Rao, J.: Dynamic resource scaling for VNF over nonstationary traffic: a learning approach. IEEE Trans. Cogn. Commun. Netw. **7**(2), 648–662 (2021)
9. Kulkarni, S.G., Zhang, W., Hwang, J., Rajagopalan, S., Ramakrishnan, K., Wood, T., et al.: NFVnice: Dynamic backpressure and scheduling for NFV service chains. IEEE/ACM Trans. Netw. **28**(2), 639–652 (2020)

Chapter 6
Conclusions and Future Research Directions

6.1 Conclusions

In this book, we study dynamic resource management solutions for embedded services in an SDN/NFV-enabled core network, with the consideration of the unique properties of CPU computing resources in a virtualized network environment, to achieve consistent QoS performance in terms of E2E delay guarantee, by adapting to network traffic dynamics in different time granularities.

We first present a delay-aware flow migration model for embedded services with average E2E delay requirements based on a simplified assumption of Poisson traffic. The packet arrival rate of the Poisson traffic is assumed stable within a time interval and varies across different time intervals. A mixed integer optimization problem is formulated, to balance between two conflicting objectives of load balancing and reconfiguration overhead reduction, under computing resource capacity constraints, average E2E delay constraints, and maximal tolerable service downtime constraints. Two solutions for the optimization problem are presented, including an optimal MIQCP solution and a low-complexity heuristic solution. With flow migration, more traffic from the services can be accommodated with average E2E delay guarantee. Numerical results demonstrate that the flow migration model achieves medium level load balancing without a significant compromise on reconfiguration overhead. The heuristic solution achieves performance comparable with the optimal MIQCP solution in terms of total cost minimization, with significant improvement on time efficiency.

In the second research problem, we remove the Poisson traffic assumption, and investigate when and how to trigger resource scaling and possible VNF migrations in a local network segment. A stricter probabilistic delay requirement is considered for QoS-aware resource demand prediction. We use traffic samples of a real-world non-stationary traffic trace in both medium timescale (20 s in simulation) and small timescale (100 ms in simulation) for resource demand prediction, based on a change point detection scheme and a GPR-based fBm traffic parameter learning scheme.

W. Zhuang, K. Qu, *Dynamic Resource Management in Service-Oriented Core Networks*, Wireless Networks, https://doi.org/10.1007/978-3-030-87136-9_6

Packet-level simulations demonstrate the effectiveness of the resource demand prediction scheme in terms of capturing the traffic burstiness in timescales larger than 100 ms. QoS satisfaction is observed for a synthesized packet arrival trace with traffic burstiness only in timescales larger than 100 ms, while occasional QoS violation is observed for the real-world packet arrival trace with traffic burstiness in even smaller time granularities such as 1 ms, especially for the more stringent QoS requirements. The outputs of resource demand prediction, including both the detected change points in time and the predicted resource demands, are applied to a dynamic VNF migration learning module based on a penalty-aware deep Q-learning algorithm. Through reinforcement learning, the patterns in the traffic trace can be captured and VNF migration decisions can be made adaptively to achieve a trade-off among load balancing, migration cost reduction, and resource overloading penalty suppression in the long run. The decision epoch length is time-varying, corresponding to the time duration of each detected stationary traffic segment. Numerical results show that the deep Q-learning algorithm achieves more training loss reduction and more penalty suppression compared with the benchmarks.

In the third research problem, we focus on a sufficiently long time duration with given VNF placement and stationary traffic statistics, and investigate delay-aware VNF scheduling for deadline-constrained services with strict QoS requirement in terms of delay violation ratio, to achieve total network utility maximization with traffic dynamics in even smaller time granularities (e.g., 100 μs–1 ms). To incorporate packet delay awareness, we use a delay-aware virtual packet processing queueing model. We also replace the QoS requirements by equivalent virtual queue stability requirements for each service. Based on Lyapunov optimization, an online distributed VNF scheduling algorithm is derived, which greedily minimizes a Lyapunov drift-plus-penalty in each time slot. The delay-aware VNF scheduling algorithm can be executed in a time slotted manner with a realistic *state-of-the-art* time slot length (e.g., 100 μs–1 ms) for CPU resource scheduling. At each NFV node, a VNF with the maximum VNF scheduling weight is scheduled. The scheduling weight for a VNF incorporates the weighted differential backlogs with the downstream VNF, the virtual queue lengths indicating the current QoS performance, and the number of urgent and non-urgent packets. The delay-aware VNF scheduling algorithm achieves asymptotically optimal total network utility with the increase of utility importance parameter, at the cost of linearly increasing average total backlogs. The effectiveness of virtual queues is verified through a significant QoS improvement achieved by the presented delay-aware VNF scheduling algorithm compared with the "Backpressure-P" benchmark. It also achieves a comparable performance with a GPS benchmark scheme under the unrealistic assumption of infinitely divisible CPU computing resources. We have also evaluated the performance of a modified VNF scheduling algorithm when packet rushing is allowed. With packet rushing, we observe performance improvements especially in the average E2E delay.

6.2 Future Research Directions

In the learning-based VNF migration scheme, the QoS provisioning and the cost minimization are addressed separately. The QoS provisioning is achieved by first predicting a QoS-aware resource demand and then allocating resources accordingly. The cost minimization is achieved by deep Q-learning based VNF migration decision. However, due to the traffic burstiness in finer time granularities than the small time interval (e.g., $100\,\mu s$) under consideration in the resource demand prediction framework, QoS violation occasionally happens especially for a very stringent QoS requirement. To provide better QoS provisioning, a potential approach is to explicitly incorporate the measured QoS performance metric as a feedback in the MDP state and make decisions accordingly to achieve QoS satisfaction in the long run. In this case, a new MDP formulation with both resource capacity constraints and QoS constraints should be investigated. The simplest way to deal with constrained MDP is reward shaping, i.e., creating a new reward as a weighted combination of the original reward and the constraint violation penalties, as done in the penalty-aware deep Q-learning approach. However, the weights in the shaped reward are typically difficult to adjust during the training process especially when the MDP model is complex with a high-dimensional state and action space and when there are multiple constraints. A potential approach to deal with constrained MDP is to apply the primal-dual method in RL, in which the weights of the constraint violation penalties are dual variables which can be updated and learned together with the VNF migration policy which corresponds to primal variables.

One RL learning agent is responsible for the adaptive migration decision of a single VNF in a local neighborhood, where the dynamics of other VNFs are treated as background traffic at the NFV nodes. In a multi-service scenario, there should be multiple such learning agents working independently at each VNF belonging to different services, which may produce suboptimal VNF migration decisions in terms of an overall cost. A direct extension of the MDP formulation with multiple VNFs may cause the curse of dimensionality issue. Multi-agent or distributed RL is a potential approach to simultaneously making decisions at multiple VNFs towards a common objective, while keeping the state and action space at each learning agent small. Considering multiple VNFs will facilitate exploiting the multiplexing gain among VNFs and improve resource utilization.

The dynamic computing resource management solutions presented in this book for dealing with traffic dynamics in core networks can be extended to radio access networks with mobile edge computing. The coordination among different types of resources (e.g., communication, computing, and caching resources) at the mobile edge should be investigated, and new issues in wireless networks should be addressed, such as end user mobility, channel quality dynamics, and transmission interference.

Appendix A
Derivation of $\tilde{\alpha}_n(k)$

Let $\eta_n^{rh}(k)$ denote a ratio between resources occupied by VNF $V_h^{(r)}$ and resource capacity of NFV node n, given by

$$\eta_n^{rh}(k) = \frac{P_n^{rh} b_r}{\mathsf{C}_n}\left(\lambda^{(r)}(k) + \frac{1}{D^{rh}(k)}\right) \mathrm{x}_n^{rh}(k). \tag{A.1}$$

VNF set $\mathcal{V}$ is divided into two subsets, i.e., $\mathcal{V} = \mathcal{V}_1 \cup \mathcal{V}_2$, where $\mathcal{V}_1 = \{(r, h) \in \mathcal{V} | \mathrm{x}_n^{rh}(k) = f_2^{(r)} = 1\}$ is a set of VNFs belonging to SFC category III on NFV node n, and $\mathcal{V}_2$ is a set of all other VNFs. Vertical delay scaling is applied to only VNFs in $\mathcal{V}_1$. Before delay scaling, resource usage at NFV node n is composed of three parts, given by

$$\eta_n(k) = \sum_{(r,h)\in\mathcal{V}_1\cup\mathcal{V}_2} \eta_n^{rh}(k) + \sum_{(r,h)\in\mathcal{V}_1\cup\mathcal{V}_2} \frac{\mathrm{w}_n(k)\mathrm{x}_n^{rh}(k)\mathtt{W}}{T_n}. \tag{A.2}$$

The ratio of resources occupied by VNFs in $\mathcal{V}_1$ before vertical delay scaling is given by

$$\sum_{(r,h)\in\mathcal{V}_1} \eta_n^{rh}(k) = \sum_{(r,h)\in\mathcal{V}_1} \frac{P_n^{rh} b_r}{\mathsf{C}_n}\left(\lambda^{(r)}(k) + \frac{1}{\mathsf{D}^{rh}(k)}\right). \tag{A.3}$$

For a vertical delay scaling by a positive coefficient $\tilde{\alpha}_n(k)$ to increase loading factor of NFV node n from $\eta_n(k)$ to η_{th}, we have the following relationship among parameters, given by

$$\eta_{th} - \sum_{(r,h)\in\mathcal{V}_2} \eta_n^{rh}(k) - \sum_{(r,h)\in\mathcal{V}_1\cup\mathcal{V}_2} \frac{\mathrm{w}_n(k)\mathrm{x}_n^{rh}(k)\mathtt{W}}{T_n}$$

W. Zhuang, K. Qu, *Dynamic Resource Management in Service-Oriented Core Networks*, Wireless Networks, https://doi.org/10.1007/978-3-030-87136-9

$$= \sum_{(r,h)\in\mathcal{V}_1} \frac{P_n^{rh} b_r}{\mathsf{C}_n} \left(\lambda^{(r)}(k) + \frac{1}{\tilde{\alpha}_n(k)\mathsf{D}^{rh}(k)} \right). \tag{A.4}$$

Subtracting (A.3) from (A.4) and arranging items, we obtain

$$\tilde{\alpha}_n(k) = \frac{\sum_{(r,h)\in\mathcal{V}_1} \frac{P_n^{rh} b_r}{\mathsf{D}^{rh}(k)}}{[\eta_{th} - \eta_n(k)]\,\mathsf{C}_n + \sum_{(r,h)\in\mathcal{V}_1} \frac{P_n^{rh} b_r}{\mathsf{D}^{rh}(k)}} \tag{A.5}$$

which is equivalent to (3.27).

Appendix B
Proof of Lemma 5.1

Let C_{max} be the maximum processing resource budget (in cycle per time slot) among all the NFV nodes in set $\mathcal{N}$, and let P_{min} be the minimum processing density (in cycle/packet) among all VNFs. We have $S_{h,m}^{(r)}(\tau) \leq S_h^{(r)}(\tau) \leq \frac{C_{max}}{P_{min}} = S_{max}$, $D_h^{(r)}(\tau) \leq Q_{h,\mathbb{U}}^{(r)}(\tau) \leq A_{max}^{(r)}$ and $\chi^{(r)}(\tau) \leq A_{max}^{(r)}$. Based on the inequality $([a-b]^+ + c)^2 \leq a^2 + b^2 + c^2 - 2a(b-c)$ for $\forall a, b, c \geq 0$, we can obtain from the queue update equation (5.2) that

$$
\begin{aligned}
&\sum_{h\in\mathcal{H}_r} \left[P_h^{(r)} q_h^{(r)}(\tau+1)\right]^2 \leq \sum_{h\in\mathcal{H}_r} \left[P_h^{(r)} q_h^{(r)}(\tau)\right]^2 \\
&- 2\sum_{h\in\mathcal{H}_r} (P_h^{(r)})^2 q_h^{(r)}(\tau) \left[S_h^{(r)}(\tau) - S_{h,\mathbb{U}}^{(r)}(\tau) - S_{h-1}^{(r)}(\tau)\mathbb{1}_{\{h>1\}} - A^{(r)}(\tau)\mathbb{1}_{\{h=1\}}\right] \\
&+ \sum_{h\in\mathcal{H}_r} (P_h^{(r)})^2 \left[S_h^{(r)}(\tau) + D_h^{(r)}(\tau)\right]^2 + \sum_{h\in\mathcal{H}_r} (P_h^{(r)})^2 \left[S_{h-1}^{(r)}(\tau)\mathbb{1}_{\{h>1\}} + A^{(r)}(\tau)\mathbb{1}_{\{h=1\}}\right]^2
\end{aligned}
\tag{B.1}
$$

where the last two terms are upper bounded by constant $\mathsf{B}_1^{(r)}$ given by

$$
\mathsf{B}_1^{(r)} = (P_{max}^{(r)})^2 \left[H_r \left(S_{max} + A_{max}^{(r)}\right)^2 + (H_r - 1)(S_{max})^2 + (A_{max}^{(r)})^2\right]. \tag{B.2}
$$

Similarly, we can obtain from the queue update equation (5.17) that

$$
\begin{aligned}
&W^{(r)}(\tau+1)^2 \leq W^{(r)}(\tau)^2 \\
&- 2W^{(r)}(\tau) \left[A^{(r)}(\tau) + \sum_{h\in\mathcal{H}_r} S_{h,\mathbb{U}}^{(r)}(\tau) - \sum_{h\in\mathcal{H}_r} Q_{h,\mathbb{U}}^{(r)}(\tau) - \bar{a}_r (1-\varepsilon_r)\right]
\end{aligned}
$$

W. Zhuang, K. Qu, *Dynamic Resource Management in Service-Oriented Core Networks*, Wireless Networks, https://doi.org/10.1007/978-3-030-87136-9

$$+\left[A^{(r)}(\tau)+\sum_{h\in\mathcal{H}_r}S_{h,\mathbb{U}}^{(r)}(\tau)\right]^2+\left[\sum_{h\in\mathcal{H}_r}Q_{h,\mathbb{U}}^{(r)}(\tau)\right]^2$$
$$+\left(\bar{a}_r\right)^2\left(1-\varepsilon_r\right)^2+2\bar{a}_r\left(1-\varepsilon_r\right)\sum_{h\in\mathcal{H}_r}Q_{h,\mathbb{U}}^{(r)}(\tau) \tag{B.3}$$

where the last four terms are upper bounded by constant $\mathsf{B}_2^{(r)}$ given by

$$\mathsf{B}_2^{(r)} = (A_{max}^{(r)}+H_r S_{max})^2+(H_r A_{max}^{(r)})^2$$
$$+\left(\bar{a}_r\right)^2\left(1-\varepsilon_r\right)^2+2\bar{a}_r\left(1-\varepsilon_r\right)H_r A_{max}^{(r)}. \tag{B.4}$$

From the queue update equation (5.18), we can obtain that

$$F^{(r)}(\tau+1)^2 \le F^{(r)}(\tau)^2$$
$$-2F^{(r)}(\tau)\left[A^{(r)}(\tau)+\sum_{h\in\mathcal{H}_r}S_{h,\mathbb{U}}^{(r)}(\tau)-\sum_{h\in\mathcal{H}_r}Q_{h,\mathbb{U}}^{(r)}(\tau)-\chi^{(r)}(\tau)\right]$$
$$+\left[A^{(r)}(\tau)+\sum_{h\in\mathcal{H}_r}S_{h,\mathbb{U}}^{(r)}(\tau)\right]^2+\left[\sum_{h\in\mathcal{H}_r}Q_{h,\mathbb{U}}^{(r)}(\tau)\right]^2$$
$$+\chi^{(r)}(\tau)^2+2\chi^{(r)}(\tau)\sum_{h\in\mathcal{H}_r}Q_{h,\mathbb{U}}^{(r)}(\tau) \tag{B.5}$$

where the last four terms are upper bounded by constant $\mathsf{B}_3^{(r)}$ given by

$$\mathsf{B}_3^{(r)} = (A_{max}^{(r)}+H_r S_{max})^2+(H_r A_{max}^{(r)})^2+(A_{max}^{(r)})^2. \tag{B.6}$$

Using the inequalities in (B.1), (B.3) and (B.5), we can obtain that

$$\Delta\left(\boldsymbol{\Theta}(\tau)\right)-\vartheta\mathbb{E}\left\{\sum_{r\in\mathcal{R}}\phi(\chi^{(r)}(\tau))|\boldsymbol{\Theta}(\tau)\right\}\le\sum_{r\in\mathcal{R}}\mathsf{B}_r$$
$$-\sum_{r\in\mathcal{R}}\mathbb{E}\left\{\sum_{h\in\mathcal{H}_r}\left(P_h^{(r)}\right)^2 q_h^{(r)}(\tau)\left[S_h^{(r)}(\tau)-S_{h,\mathbb{U}}^{(r)}(\tau)\right.\right.$$
$$\left.\left.-S_{h-1}^{(r)}(\tau)\mathbb{1}_{\{h>1\}}-A^{(r)}(\tau)\mathbb{1}_{\{h=1\}}\right]|\boldsymbol{\Theta}(\tau)\right\}$$

$$
\begin{aligned}
&- \sum_{r\in\mathcal{R}} \mathbb{E}\left\{\left(\frac{\varphi}{\bar{a}_r}\right)^2 W^{(r)}(\tau)\left[A^{(r)}(\tau) + \sum_{h\in\mathcal{H}_r} S_{h,\mathbb{U}}^{(r)}(\tau)\right.\right.\\
&\qquad\qquad \left.\left. - \sum_{h\in\mathcal{H}_r} Q_{h,\mathbb{U}}^{(r)}(\tau) - \bar{a}_r\,(1-\varepsilon_r)\right] |\boldsymbol{\Theta}(\tau)\right\}\\
&- \sum_{r\in\mathcal{R}} \mathbb{E}\left\{(P_r)^2 F^{(r)}(\tau)\left[A^{(r)}(\tau) + \sum_{h\in\mathcal{H}_r} S_{h,\mathbb{U}}^{(r)}(\tau)\right.\right.\\
&\qquad\qquad \left.\left. - \sum_{h\in\mathcal{H}_r} Q_{h,\mathbb{U}}^{(r)}(\tau) - \chi^{(r)}(\tau)\right] |\boldsymbol{\Theta}(\tau)\right\}\\
&+ \sum_{r\in\mathcal{R}} \mathbb{E}\left\{(P_r)^2 \chi^{(r)}(\tau) \sum_{h\in\mathcal{H}_r} Q_{h,\mathbb{U}}^{(r)}(\tau)|\boldsymbol{\Theta}(\tau)\right\}\\
&- \vartheta\mathbb{E}\left\{\sum_{r\in\mathcal{R}} \phi(\chi^{(r)}(\tau))|\boldsymbol{\Theta}(\tau)\right\}
\end{aligned} \tag{B.7}
$$

where B_r is a constant for service $r \in \mathcal{R}$, given by

$$
\mathsf{B}_r = \frac{1}{2}\left[\mathsf{B}_1^{(r)} + \left(\frac{\varphi}{\bar{a}_r}\right)^2 \mathsf{B}_2^{(r)} + (P_r)^2\,\mathsf{B}_3^{(r)}\right]. \tag{B.8}
$$

The inequality in (B.7) can be rewritten as (5.22) in Chap. 5.

Appendix C
VNF Scheduling Algorithm with Packet Rushing

When packet rushing is allowed, the extra packets processed by VNF $V_h^{(r)}$ are referred to as rushing packets for VNF $V_h^{(r)}$. The packets which are originally in the queue are referred to as non-rushing packets. The number of non-rushing packets processed at VNF $V_h^{(r)}$ during time slot τ is $S_h^{(r)}(\tau)$. Since it is possible that a packet can be processed by several consecutive VNFs during one time slot, the rushing packets processed by VNF $V_h^{(r)}$ can include both rushing and non-rushing packets processed by upstream VNF $V_{h-1}^{(r)}$.

Packet Rushing Analysis

Assume that the VNF scheduling variables, i.e., $\{z_h^{(r)}(\tau), \forall h \in \mathcal{H}_r\}$, and the number of non-rushing packets processed at each VNF, i.e., $\{S_h^{(r)}(\tau), \forall h \in \mathcal{H}_r\}$, are given for service $r \in \mathcal{R}$ during time slot τ. We analyze the number of rushing packets processed by each VNF of service $r \in \mathcal{R}$ during time slot τ, denoted by $R_h^{(r)}(\tau)$ for VNF $V_h^{(r)}$. The VNFs of service $r \in \mathcal{R}$ are classified into three subsets for time slot τ, given by

$$\mathcal{H}_1^{(r)}(\tau) = \{h \in \mathcal{H}_r | z_h^{(r)}(\tau) = 0\}, \quad r \in \mathcal{R} \tag{C.1}$$

$$\mathcal{H}_2^{(r)}(\tau) = \{h \in \mathcal{H}_r \backslash \{1\} | z_h^{(r)}(\tau) = 1, z_{h-1}^{(r)}(\tau) = 1, S_h^{(r)}(\tau) < \left\lfloor \frac{\sum_{n \in \mathcal{N}} x_n^{rh} C_n}{P_h^{(r)}} \right\rfloor\}, \; r \in \mathcal{R} \tag{C.2}$$

$$\mathcal{H}_3^{(r)}(\tau) = \mathcal{H}_r \backslash \left(\mathcal{H}_1^{(r)}(\tau) \cup \mathcal{H}_2^{(r)}(\tau)\right), \quad r \in \mathcal{R}. \tag{C.3}$$

W. Zhuang, K. Qu, *Dynamic Resource Management in Service-Oriented Core Networks*, Wireless Networks, https://doi.org/10.1007/978-3-030-87136-9

For service $r \in \mathcal{R}$, subset $\mathcal{H}_1^{(r)}(\tau)$ includes all the unscheduled VNFs during time slot τ, subset $\mathcal{H}_2^{(r)}(\tau)$ includes all the scheduled VNFs where there is packet rushing opportunity, i.e., the scheduled unsaturated VNFs whose upstream VNF is also scheduled, and subset $\mathcal{H}_3^{(r)}(\tau)$ includes all the scheduled VNFs where there is no packet rushing. Intuitively, the overall packet rushing opportunity is higher, if we have more VNFs in subset $\mathcal{H}_2^{(r)}(\tau)$ with more residual resources. Let $\tilde{S}_h^{(r)}(\tau)$ be the actual number of packets processed by VNF $V_h^{(r)}$ during time slot τ with the consideration of packet rushing, given by

$$\tilde{S}_h^{(r)}(\tau) = S_h^{(r)}(\tau) + R_h^{(r)}(\tau), \quad \forall r \in \mathcal{R},\ h \in \mathcal{H}_r. \tag{C.4}$$

We have $\tilde{S}_h^{(r)}(\tau) = S_h^{(r)}(\tau), \forall h \in \mathcal{H}_1^{(r)}(\tau) \cup \mathcal{H}_3^{(r)}(\tau)$ and $\tilde{S}_h^{(r)}(\tau) \geq S_h^{(r)}(\tau), \forall h \in \mathcal{H}_2^{(r)}(\tau)$ for service $r \in \mathcal{R}$ during time slot τ.

We consider a finite timeline, $\mathrm{t} \in [0, T]$ starting at the beginning of time slot τ and ending at the end of time slot τ, where T is the time slot length in second. Assume that the scheduled VNFs during time slot τ start to process the first packet in their queues at time instant $t = 0$. Let $\upsilon_{h,max}^{(r)}$ denote the maximum packet processing rate (in packet/s) for VNF $V_h^{(r)}$, given by

$$\upsilon_{h,max}^{(r)} = \frac{\sum_{n\in\mathcal{N}} \mathrm{x}_n^{rh} C_n}{P_h^{(r)} T}, \quad \forall r \in \mathcal{R},\ \forall h \in \mathcal{H}_r. \tag{C.5}$$

For VNF $V_h^{(r)}$ in subset $\mathcal{H}_2^{(r)}(\tau)$, there are $S_h^{(r)}(\tau)$ non-rushing packets being processed first in the maximum packet processing rate $\upsilon_{h,max}^{(r)}$. Then, the rushing packets from the upstream VNF $V_{h-1}^{(r)}$ start to be processed at VNF $V_h^{(r)}$. However, the actual processing rate for the rushing packets depend on the packet processing rate of either VNF $V_h^{(r)}$ or VNF $V_{h-1}^{(r)}$ in different conditions. Let $\upsilon_h^{(r)}(\mathrm{t})$ denote the actual packet processing rate (in packet/s) for VNF $V_h^{(r)}$ at time $\mathrm{t} \in [0, T]$, with $\upsilon_h^{(r)}(0) = \upsilon_{h,max}^{(r)}$. There are three cases for $\upsilon_h^{(r)}(\mathrm{t})$, depending on the relationships among $\upsilon_{h,max}^{(r)}$, $\upsilon_{h-1}^{(r)}(\mathrm{t})$, $S_h^{(r)}(\tau)$ and T. Figures C.1 and C.2 illustrate the three cases for $\upsilon_h^{(r)}(\mathrm{t})$, with $\upsilon_{h-1}^{(r)}(\mathrm{t})$ being either a constant or a decreasing step function with t, respectively.

- Case 1: If VNF $V_h^{(r)}$ cannot process packets faster than VNF $V_{h-1}^{(r)}$ within time duration T, i.e., $\upsilon_{h,max}^{(r)} \leq \min_{\mathrm{t}\in[0,T]} \upsilon_{h-1}^{(r)}(\mathrm{t})$, we have $\upsilon_h^{(r)}(\mathrm{t}) = \upsilon_{h,max}^{(r)}, \forall \mathrm{t} \in [0, T]$, as illustrated in Figures C.1a and C.2a.
- Case 2: Under the condition that $\upsilon_{h,max}^{(r)} > \min_{\mathrm{t}\in[0,T]} \upsilon_{h-1}^{(r)}(\mathrm{t})$, if the number of packets processed by VNF $V_h^{(r)}$ in the maximum processing rate $\upsilon_{h,max}^{(r)}$ within time duration T does not exceed the number of packets processed by VNF $V_{h-1}^{(r)}$

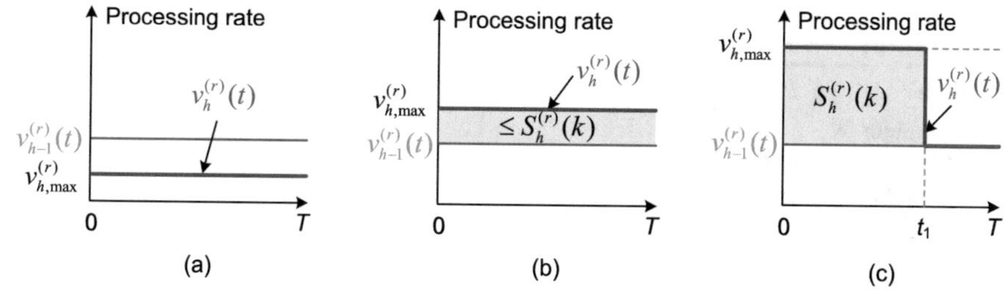

Fig. C.1 An illustration of actual packet processing rate $\upsilon_h^{(r)}(\mathrm{t})$ if $\upsilon_{h-1}^{(r)}(\mathrm{t})$ is a constant

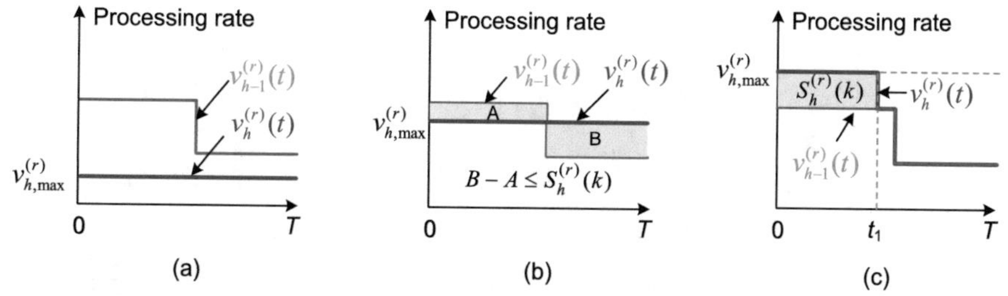

Fig. C.2 An illustration of actual packet processing rate $\upsilon_h^{(r)}(\mathrm{t})$ if $\upsilon_{h-1}^{(r)}(\mathrm{t})$ is a decreasing step function

within time duration T by more than $S_h^{(r)}(\tau)$, i.e., $\upsilon_{h,max}^{(r)}T - \int_0^T \upsilon_{h-1}^{(r)}(\mathrm{t})d\mathrm{t} \leq S_h^{(r)}(\tau)$, the actual processing rate for both the non-rushing and rushing packets at VNF $V_h^{(r)}$ is equal to $\upsilon_{h,max}^{(r)}$ within time duration T, as illustrated in Figures C.1b and C.2b.

- Case 3: Under the condition that $\upsilon_{h,max}^{(r)} > \min_{\mathrm{t}\in[0,T]} \upsilon_{h-1}^{(r)}(\mathrm{t})$, if VNF $V_h^{(r)}$ can process more packets than VNF $V_{h-1}^{(r)}$ by $S_h^{(r)}(\tau)$ at a certain time instant $\mathrm{t}_1 < T$, i.e., $\upsilon_{h,max}^{(r)}\mathrm{t}_1 - \int_0^{\mathrm{t}_1} \upsilon_{h-1}^{(r)}(\mathrm{t})d\mathrm{t} = S_h^{(r)}(\tau)$, the actual packet processing rate for VNF $V_h^{(r)}$ is equal to that of the upstream VNF $V_{h-1}^{(r)}$ after time t_1, as illustrated in Figures C.1c and C.2c. We refer to time instant t_1 as transition time instant.

We see that the actual packet processing rate $\upsilon_h^{(r)}(\mathrm{t})$ is either a constant or a decreasing step function with t, no matter $\upsilon_{h-1}^{(r)}(\mathrm{t})$ is a constant or a decreasing step function with t. The number of rushing packets processed by VNF $V_h^{(r)}$ in subset $\mathcal{H}_2^{(r)}(\tau)$ during time slot τ, i.e., $R_h^{(r)}(\tau)$, is limited by the actual number of packets processed by the upstream VNF $V_{h-1}^{(r)}$ during time slot τ and the maximum number of extra packets that VNF $V_h^{(r)}$ can process in the actual packet processing rate $\upsilon_h^{(r)}(\mathrm{t})$ within time duration T, represented by

$$R_h^{(r)}(\tau) = \min\left(\tilde{S}_{h-1}^{(r)}(\tau), \int_0^T \upsilon_h^{(r)}(\mathrm{t})d\mathrm{t} - S_h^{(r)}(\tau)\right). \tag{C.6}$$

Algorithm: Packet rushing analysis for service $r \in \mathcal{R}$ during time slot τ

Input: $\{S_h^{(r)}(\tau), h \in \mathcal{H}_r\}$, sets $\mathcal{H}_1^{(r)}(\tau), \mathcal{H}_2^{(r)}(\tau), \mathcal{H}_3^{(r)}(\tau)$.
Initialize: $\{R_h^{(r)}(\tau) = 0, h \in \mathcal{H}_r\}$.
for $h = 1, \cdots, H_r$ **do**
  if $h \in \mathcal{H}_3^{(r)}(\tau)$, **then**
    $\boldsymbol{v}_h^{(r)} = v_{h,max}^{(r)}, \mathbf{t}_h^{(r)} = [0, T]^{\mathsf{T}}$.
  if $h \in \mathcal{H}_2^{(r)}(\tau)$ **then**
    if $v_{h,max}^{(r)} \leq \min\left(\boldsymbol{v}_{h-1}^{(r)}\right)$ **then**
      $\boldsymbol{v}_h^{(r)} = v_{h,max}^{(r)}, \mathbf{t}_h^{(r)} = [0, T]^{\mathsf{T}}$.
    else
      if $v_{h,max}^{(r)} T - \boldsymbol{v}_{h-1}^{(r)\,\mathsf{T}} \cdot \left(\mathbf{t}_{h-1}^{(r)}\left[2 : |\mathbf{t}_{h-1}^{(r)}|\right] - \mathbf{t}_{h-1}^{(r)}\left[1 : |\boldsymbol{v}_{h-1}^{(r)}|\right]\right) \leq S_h^{(r)}(\tau)$ **then**
        $\boldsymbol{v}_h^{(r)} = v_{h,max}^{(r)}, \mathbf{t}_h^{(r)} = [0, T]^{\mathsf{T}}$.
      else
        $\boldsymbol{\delta} = \left(v_{h,max}^{(r)} \mathbf{e}_{|\boldsymbol{v}_{h-1}^{(r)}|} - \boldsymbol{v}_{h-1}^{(r)}\right) \circ \left(\mathbf{t}_{h-1}^{(r)}\left[2 : |\mathbf{t}_{h-1}^{(r)}|\right] - \mathbf{t}_{h-1}^{(r)}\left[1 : |\boldsymbol{v}_{h-1}^{(r)}|\right]\right)$.
        Find $\mathtt{j}_0$ with $\sum_{\mathtt{j} \leq \mathtt{j}_0} \boldsymbol{\delta}(\mathtt{j}) > S_h^{(r)}(\tau) \geq \sum_{\mathtt{j} < \mathtt{j}_0} \boldsymbol{\delta}(\mathtt{j})$.
        Calculate transition time instant $\mathtt{t}_1 = \mathbf{t}_{h-1}^{(r)}(\mathtt{j}_0 + 1) - \frac{\sum_{\mathtt{j} \leq \mathtt{j}_0} \boldsymbol{\delta}(\mathtt{j}) - S_h^{(r)}(\tau)}{v_{h,max}^{(r)} - \boldsymbol{v}_{h-1}^{(r)}(\mathtt{j}_0)}$.
        $\boldsymbol{v}_h^{(r)} = \left[v_{h,max}^{(r)}, \boldsymbol{v}_{h-1}^{(r)}\left[\mathtt{j}_0 : |\boldsymbol{v}_{h-1}^{(r)}|\right]^{\mathsf{T}}\right]^{\mathsf{T}}$.
        $\mathbf{t}_h^{(r)} = \left[0, \mathtt{t}_1, \mathbf{t}_{h-1}^{(r)}\left[\mathtt{j}_0 + 1 : |\mathbf{t}_{h-1}^{(r)}|\right]^{\mathsf{T}}\right]^{\mathsf{T}}$.
    $R_h^{(r)}(\tau) = \min\left(\tilde{S}_{h-1}^{(r)}(\tau), \left\lfloor \boldsymbol{v}_h^{(r)\mathsf{T}} \cdot \left(\mathbf{t}_h^{(r)}\left[2 : |\mathbf{t}_h^{(r)}|\right] - \mathbf{t}_h^{(r)}\left[1 : |\boldsymbol{v}_h^{(r)}|\right]\right)\right\rfloor - S_h^{(r)}(\tau)\right)$.
  Calculate $\tilde{S}_h^{(r)}(\tau)$ according to (C.4).
Output: $\{R_h^{(r)}(\tau), h \in \mathcal{H}_r\}, \{\tilde{S}_h^{(r)}(\tau), h \in \mathcal{H}_r\}$.

For service $r \in \mathcal{R}$, packet rushing analysis is performed iteratively for VNFs from source to destination, with a pseudo code presented in an Algorithm. In the algorithm, we use two vectors, $\boldsymbol{v}_h^{(r)}$ and $\mathbf{t}_h^{(r)}$, to represent function $v_h^{(r)}(\mathtt{t}), \mathtt{t} \in [0, T]$. Let $\boldsymbol{v}_h^{(r)}$ be a vector of actual packet processing rates for VNF $V_h^{(r)}$, e.g., $\boldsymbol{v}_h^{(r)} = [100, 80, 60]^{\mathsf{T}}$, and let $\mathbf{t}_h^{(r)}$ be a vector of time boundaries between the actual packet processing rates, e.g., $\mathbf{t}_h^{(r)} = [0, \frac{T}{2}, \frac{3T}{4}, T]^{\mathsf{T}}$, where superscript T denotes the transpose operator. The dimension of vector $\mathbf{t}_h^{(r)}$ is larger than that of vector $\boldsymbol{v}_h^{(r)}$ by 1, i.e., $|\mathbf{t}_h^{(r)}| = |\boldsymbol{v}_h^{(r)}| + 1$. The relationship between $v_h^{(r)}(\mathtt{t})$ and the two vectors is given by

$$v_h^{(r)}(\mathtt{t}) = \boldsymbol{v}_h^{(r)}(\mathtt{j}), \quad \text{if } \mathbf{t}_h^{(r)}(\mathtt{j}) \leq t < \mathbf{t}_h^{(r)}(\mathtt{j}+1) \tag{C.7}$$

where $\mathtt{j}$ is the index of the $\mathtt{j}$-th ($\mathtt{j} \leq |\boldsymbol{v}_h^{(r)}|$) element in vector $\boldsymbol{v}_h^{(r)}$ or $\mathbf{t}_h^{(r)}$. Then, we have

$$\int_0^T \upsilon_h^{(r)}(\mathrm{t})d\mathrm{t} = \boldsymbol{\upsilon}_h^{(r)\mathsf{T}} \cdot \left(\mathbf{t}_h^{(r)}\left[2 : |\mathbf{t}_h^{(r)}|\right] - \mathbf{t}_h^{(r)}\left[1 : |\boldsymbol{\upsilon}_h^{(r)}|\right]\right). \tag{C.8}$$

In the packet rushing analysis algorithm, Lines 7–8 correspond to Case 1 for $\upsilon_h^{(r)}(\mathrm{t})$ of VNF $V_h^{(r)}$ in subset $\mathcal{H}_2^{(r)}(\tau)$, Lines 10–11 correspond to Case 2, and Lines 12–17 correspond to Case 3. In Line 13, $\mathbf{e}_{|\boldsymbol{\upsilon}_{h-1}^{(r)}|}$ is a vector of length $|\boldsymbol{\upsilon}_{h-1}^{(r)}|$ with every element equal to 1, and $\circ$ represents the element-wise product operation between two vectors. Index $\mathtt{j}_0$ in Line 14 is the smallest index satisfying $\upsilon_{h,max}^{(r)}\mathbf{t}_{h-1}^{(r)}(\mathtt{j}_0 + 1) - \int_0^{\mathbf{t}_{h-1}^{(r)}(\mathtt{j}_0+1)} \upsilon_{h-1}^{(r)}(\mathrm{t})d\mathrm{t} > S_h^{(r)}(\tau)$.

Modified VNF Scheduling Algorithm

By taking advantage of packet rushing, a packet with residual lifetime $1 \le m < H_r - h + 1$ at VNF $V_h^{(r)}$ has opportunity to be successfully delivered before expiry, and a packet with $m = M_r$ at VNF $V_1^{(r)}$ has opportunity to rush through all the VNFs in a service to the egress edge switch in one time slot. Hence, the set of packet residual lifetime at VNF $V_h^{(r)}$, denoted by $\tilde{\mathcal{M}}_h^{(r)}$, is modified to

$$\tilde{\mathcal{M}}_h^{(r)} = \{1, \cdots, M_r\}, \quad \forall r \in \mathcal{R},\ h \in \mathcal{H}_r. \tag{C.9}$$

However, in the worst case, a packet with residual lifetime $1 \le m < H_r - h + 1$ at VNF $V_h^{(r)}$ cannot be timely delivered if there is no packet rushing opportunity. As discussed in Sect. C, whether packet rushing can happen at a certain VNF or not during a given time slot is unknown until the VNF scheduling and packet processing decisions for the time slot are given. Hence, the VNF scheduling algorithm makes worst-case decisions under the assumption of no packet rushing in each time slot, and determines the number of non-rushing packets processed at each VNF. In the worst case, all the packets with residual lifetime of $1 \le m \le H_r - h + 1$ at VNF $V_h^{(r)}$ are urgent packets since they will be eventually dropped without any packet rushing opportunity if not processed in the current time slot, and other packets are non-urgent packets. Accordingly, we have $Q_{h,\mathbb{U}}^{(r)}(\tau) = \sum_{m=1}^{H_r-h+1} Q_{h,m}^{(r)}(\tau)$ and $S_{h,\mathbb{U}}^{(r)}(\tau) = \sum_{m=1}^{H_r-h+1} S_{h,m}^{(r)}(\tau)$.

The modified VNF scheduling algorithm is derived based on the worst-case Lyapunov drift-plus-penalty, in which the physical and virtual queue lengths are updated according to (5.2), (5.17) and (5.18) with the new definitions of $Q_{h,\mathbb{U}}^{(r)}(\tau)$ and $S_{h,\mathbb{U}}^{(r)}(\tau)$. We use $D_h^{(r)}(\tau) = Q_{h,\mathbb{U}}^{(r)}(\tau) - S_{h,\mathbb{U}}^{(r)}(\tau)$ as the worst-case number of dropped packets at VNF $V_h^{(r)}$ in (5.2). In the modified algorithm, the auxiliary variable decision is the same as that without packet rushing, except for using the new definition of $Q_{h,\mathbb{U}}^{(r)}(\tau)$. For VNF scheduling and packet processing, we consider an FCFS prioritization principle for the packets with different residual lifetime at the scheduled VNFs. If VNF $V_h^{(r)}$ is scheduled, the number of non-rushing packets with residual lifetime $m \in \tilde{\mathcal{M}}_h^{(r)}$ that are processed at VNF $V_h^{(r)}$ during time slot τ, i.e., $\hat{S}_{h,m}^{(r)}(\tau)$, is given by (5.8) where $m_0 \in \tilde{\mathcal{M}}_h^{(r)}$ satisfies (5.9). Then, the optimal VNF scheduling and packet processing decisions are made based on a modified VNF scheduling weight, $\tilde{\omega}_h^{(r)}(\tau)$, given by

$$\tilde{\omega}_h^{(r)}(\tau) = \omega_{h,\mathbb{U}}^{(r)}(\tau) \sum_{m=1}^{H_r-h+1} \hat{S}_{h,m}^{(r)}(\tau) + \omega_{h,\mathbb{N}}^{(r)}(\tau) \sum_{m>H_r-h+1} \hat{S}_{h,m}^{(r)}(\tau),$$
$$\forall r \in \mathcal{R},\ h \in \mathcal{H}_r. \tag{C.10}$$

After all decisions for time slot τ are made using the modified VNF scheduling algorithm, packet rushing analysis is performed for each service. Although the modified algorithm is derived based on the worst-case queue length updates, the true queue lengths can be updated at the end of time slot τ. The true physical queue length evolution equations are

$$q_h^{(r)}(\tau+1) = \left[q_h^{(r)}(\tau) - S_h^{(r)}(\tau) - D_h^{(r)}(\tau)\right]^+ + \left[\tilde{S}_{h-1}^{(r)}(\tau) - R_h^{(r)}(\tau)\right] \mathbb{1}\{h > 1\},$$
$$+ A^{(r)}(\tau)\mathbb{1}\{h = 1\}, \quad \forall r \in \mathcal{R},\ \forall h \in \mathcal{H}_r \tag{C.11}$$

where $D_h^{(r)}(\tau)$ is updated as $D_h^{(r)}(\tau) = \left[Q_{h,1}^{(r)}(\tau) - S_{h,1}^{(r)}(\tau)\right]^+$, since only the packets with residual lifetime $m = 1$ are actually dropped at VNF $V_h^{(r)}$ if they are not processed. Correspondingly, the true virtual queue length evolution equations in (5.17) and (5.18) are updated with $Q_{h,\mathbb{U}}^{(r)}(\tau) = Q_{h,1}^{(r)}(\tau)$ and $S_{h,\mathbb{U}}^{(r)}(\tau) = S_{h,1}^{(r)}(\tau)$. Let $R_{h,m}^{(r)}(\tau)$ and $\tilde{S}_{h,m}^{(r)}(\tau)$ be the number of rushing packets with residual lifetime $m \in \tilde{\mathcal{M}}_h^{(r)}$ and the actual total number of packets with residual lifetime $m \in \tilde{\mathcal{M}}_h^{(r)}$ that are processed at VNF $V_h^{(r)}$ during time slot τ respectively, with $\tilde{S}_{h,m}^{(r)}(\tau) = S_{h,m}^{(r)}(\tau) + R_{h,m}^{(r)}(\tau)$. For VNF $V_h^{(r)}$ with $R_h^{(r)}(\tau) = 0$, we have $R_{h,m}^{(r)}(\tau) = 0$ and $\tilde{S}_{h,m}^{(r)}(\tau) = S_{h,m}^{(r)}(\tau)$ for $\forall m \in \tilde{\mathcal{M}}_h^{(r)}$. For VNF $V_h^{(r)}$ with $R_h^{(r)}(\tau) = \sum_{m\in\tilde{\mathcal{M}}_h^{(r)}} R_{h,m}^{(r)}(\tau) > 0$, the rushing packets arrive at VNF $V_h^{(r)}$ in ascending order of packet residual lifetime. Thus, $R_{h,m}^{(r)}(\tau)$ is given by

$$R_{h,m}^{(r)}(\tau) = \begin{cases} \tilde{S}_{h-1,m}^{(r)}(\tau), & \text{if } 1 \le m < m_1 \\ R_h^{(r)}(\tau) - \sum_{m<m_1} \tilde{S}_{h-1,m}^{(r)}(\tau), & \text{if } m = m_1 \\ 0, & \text{otherwise} \end{cases} \tag{C.12}$$

where $m_1 \in \tilde{\mathcal{M}}_h^{(r)}$ satisfies $\sum_{m=1}^{m_1} \tilde{S}_{h-1,m}^{(r)}(\tau) \geq R_h^{(r)}(\tau) > \sum_{m<m_1} \tilde{S}_{h-1,m}^{(r)}(\tau)$. With packet rushing, the queueing dynamics of the delay-aware virtual packet processing queues are updated as

$$
\begin{aligned}
&Q_{1,M_r}^{(r)}(\tau+1) = A^{(r)}(\tau), && \forall r \in \mathcal{R} \\
&Q_{1,m}^{(r)}(\tau+1) = \left[Q_{1,m+1}^{(r)}(\tau) - S_{1,m+1}^{(r)}(\tau)\right]^+, && \forall r \in \mathcal{R}, \quad \forall m \in \tilde{\mathcal{M}}_1^{(r)} \backslash \{M_r\} \\
&Q_{h,M_r}^{(r)}(\tau+1) = 0, && \forall r \in \mathcal{R}, \quad \forall h \in \mathcal{H}_r \backslash \{1\} \\
&Q_{h,M_r-1}^{(r)}(\tau+1) = \tilde{S}_{h-1,M_r}^{(r)}(\tau) - R_{h,M_r}^{(r)}(\tau), && \forall r \in \mathcal{R}, \quad \forall h \in \mathcal{H}_r \backslash \{1\} \\
&Q_{h,m}^{(r)}(\tau+1) = \left[Q_{h,m+1}^{(r)}(\tau) - S_{h,m+1}^{(r)}(\tau)\right]^+ \\
&\qquad\qquad + [\tilde{S}_{h-1,m+1}^{(r)}(\tau) - R_{h,m+1}^{(r)}(\tau)], && \forall r \in \mathcal{R}, \quad \forall h \in \mathcal{H}_r \backslash \{1\}, \\
& && \forall m \in \tilde{\mathcal{M}}_h^{(r)} \backslash \{M_r - 1, M_r\}.
\end{aligned}
\tag{C.13}
$$

Printed in the United States
by Baker & Taylor Publisher Services